Calendars, Symbols, and Orientations: Legacies of Astronomy in Culture

A mother teaching her daughter and a father his son how to use a calendar staff. From *Historia de gentibus septentrionalibus* by Olaus Magnus, Rome, 1555, book 4, chapter 6 (p. 203 in the 1996 English translation by P. G. Foote). See the articles by Hallonquist and Henriksson in this volume.

Calendars, Symbols, and Orientations: Legacies of Astronomy in Culture

Proceedings of the 9th annual meeting of

the European Society for Astronomy in Culture (SEAC)

The Old Observatory, Stockholm, 27-30 August 2001

Edited by Mary Blomberg, Peter E. Blomberg, and Göran Henriksson

Uppsala 2003

Uppsala Astronomical Observatory Report No. 59

Cover picture: The Old Observatory at Uppsala

Frontispiece from *Historia de gentibus septentrionalibus* by Olaus Magnus, Rome, 1555, book 4, chapter 6.

Published with the aid of grants from the Bank of Sweden Tercentenary Foundation and the Gunvor and Josef Anér's Foundation.

First published 2003

ISBN 91-506-1674-9

Printed by the University printers, Uppsala University

Contents

Preface

The present volume contains twenty-four peer-reviewed articles by scholars in the fields of archaeology, astronomy, classics, engineering, European languages, folklore, mathematics, and medieval studies, presented at the ninth annual conference of the Société Européenne pour l'Astronomie dans la Culture (SEAC) in Stockholm at the Old Observatory in 2001. The common denominator of these scholars is their interest in the influence of astronomy on other cultural manifestations such as architecture and the other arts, calendars, folk customs, and the history of ideas. The range of subjects is a vast one, and the term *archaeoastronomy* is frequently used nowadays to cover it. The keynote speech by Bengt Gustafsson, professor of theoretical astrophysics at Uppsala University, was a eulogy to such interdisciplinary interests, to the "virtue of looking in another direction" and the benefits which have accrued to society as a result.

The decision to hold the conference in Stockholm provided the welcomed opportunity for the presentation of Swedish artefacts and monuments that reflect ancient astronomical knowledge, knowledge that can be traced back to as early as the Neolithic Period. A main reason for the selection of venues for the SEAC conferences in the different countries of Europe is the concern on the part of the society to provide this opportunity. For a brief account of the history of the society and its conferences see the preface to the meeting in Dublin in 1998; a list of earlier publications appears on the back cover of this issue. More information is available at SEAC's web site http://www.iac.es/seac/seac.html.

The contributions to these *Proceedings* have been arranged according to subject matter rather than to geographical areas or chronological periods, and the topics are familiar from earlier conferences: methodology, calendars, orientations, history of the constellations, and the influence of astronomy on art, literature, folklore, and mythology. As always, not all articles fit neatly into the categories chosen, or several could have been placed in more than one. We hope that there is no uncomfortable jarring of sensibilities in the choices made.

Focusing attention on the central role of good methodology and theoretical awareness has been a concern from the beginning to the organisers of the SEAC meetings. The paper on archaeoastronomical theory by Stanislaw Iwaniszewski presents in a lucid fashion some of the problems over which many of us agonize in our work. In line with this interest, a roundtable discussion on "Theoretical and methodological considerations in archaeoastronomy and astronomy in culture" was held for the first time, having been organised by Roslyn Frank and with contributions from Juan Antonio Belmonte, Stanislaw Iwaniszewski, Frank Prendegast, and Clive Ruggles. The decision was made to continue this much appreciated innovation at later conferences and to try and include the contributions and following discussion in the *Proceedings*.

The range in time and space covered by the other papers reaches back to Palaeolithic cave art (Michael Rappenglück) and outwards to the further reaches of Siberia (Leonid Marsadolov). It is perhaps well to point out that 'European' in the case of SEAC includes scholars of all continents studying astronomy in the cultures of Europe and European scholars studying astronomy in the cultures of all continents. The article by Robert Hannah and Marina Moss is an example of the former, and there are four examples of the latter in the papers of Juan Antonio Belmonte, César Estaban, Yves and Christine Gauthier, and Arkadiusz Sołtysiak.

The choice of Sweden for the ninth conference was especially fortuitous, as publications of its archaeoastronomical material in languages other than the Scandinavian are rare. For example the remarkable runic calendar staffs (frontispiece) are little known outside Sweden, and we are fortunate in having Sven-Göran Hallonquist present a paper on these remarkable instruments. They were used in Sweden from the Middle Ages, in the first place as an aid in keeping track of the Christian ritual calendar over the years, but they also contained other information. Hallonquist's contribution had the extra advantage of providing a background for the visit to the University Museum in the Gustavianum at Uppsala, which has an exceptionally fine collection of such staffs. Göran Henriksson's article on the 'royal' burial mounds at Old Uppsala also served to deepen the experience of the visit to that site, where the author, together with Professor Ola Kyhlberg of the Department of Archaeology and Ancient History at Uppsala University, gave a highly qualified guided tour. The astronomical significance of these monuments lies in the fact that they are aligned to sunset on the date of 8 February, a fact that could have been used to regulate the important eight-year cycle of the pagan sacrificial period and the later assembly and market at Old Uppsala, which took place at the same time after the Christianisation of Sweden.

Orientation studies continue to have a central place at SEAC conferences, as exemplified by the contributions of César Gonzáles-García and Lourdes Costa-Ferrer, Leonid Marsadolov, Mary Blomberg and Göran Henriksson, César Esteban, Yves and Christine Gauthier, and Florin Stanescu. However the emphasis is shifting more and more from the presentation of the data, with only brief discussion of the possible reasons for the orientations, to the interpretation of these data as manifestations of the culture in which they occur. It is also gratifying to note that statistical evaluations and attention to the broader archaeological context are more routinely a part of such studies.

A number of papers deal with calendar traditions, for example the contributions of Hallonquist and Henriksson mentioned above. Evidence gathered primarily from the folklore of eastern Europe is presented in the papers of Jonas Vaiškūnas and Vessalina Koleva, whereas orientations play a larger role in the contribution of Stanislaw Iwaniszewski on Stonehenge and that of Dmitry Zdanovich and Andrey Kirillov.

The history of the constellations has recently prompted lively discussions on the HASTRO site for the history of astronomy, and this interest is reflected in several articles—those by Juan Antonio Belmonte, Peter E. Blomberg, Robert Hannah and Marina Moss, Michael Rappenglück, Sergey Zhitomirsky, and Izold Pustylnik.

The influence and reflections of astronomical knowledge on art, mythology, and literature are the subjects dealt with in the papers by Barbara Rappenglück, Rafal Perkowski, Arkadiusz Sołtysiak, and Emília Pásztor.

The editors wish to express their deep gratitude to the following institutions and individuals that helped to make the conference possible and successful: the Swedish Research Council, the Bank of Sweden Tercentenary Foundation, and the Gunvor and Josef Anér's Foundation for financial support; Bengt Gustafsson, Ola Kyhlberg, and Sven-Göran Hallonquist for kindly agreeing to give papers on especially interesting topics for the conference; Clive Ruggles for his lecture on Stonehenge at the Department of Archaeology and Ancient History at Uppsala; Inga Elmqvist, curator at the Observatory Museum, for allowing us to have the conference in the beautiful and historical building more popularly known as the Old Observatory; the city of Stockholm for generously hosting a buffet dinner in the City Hall and to the Lord Major and our guide of the building for making the evening memorable; Ola Kyhlberg and Wladyslaw Duczko for guiding us at Old Uppsala; Christina Risberg, first curator at the Uppsala University Museum in the Gustavianum, and Sven-Göran Hallonquist for making available to us runic calendar staffs not normally on display in the University collection; and Brita Alroth and Mats Cullhed for arranging the buffet dinner following our visit. We would also like to thank Claes-Ingvar Lagerkvist, chairman of the Department of Astronomy and Space Physics at Uppsala University, for accepting this volume for the Uppsala Astronomical Observatory Report series.

We were very fortunate in being able to conclude the conference with an excursion to some of the most interesting archaeoastronomical sites in Sweden: the passage graves and field of menhirs of Falbygden and the rock carvings of the Norrköping area. We owe a special debt of gratitude to Ann-Charlotte Hertz, first city antiquarian of Norrköping, for guiding us at the rock-carving sites of her district and to Göran Henriksson for demonstrating his interpretation of the astronomical content of these monuments and the orientations of the passage graves. The excursion also included visits to Rökstenen in Östergötland, Sweden's largest rune stone, and the huge ship setting at Askeberga.

We are especially grateful to the anonymous reviewers who so generously offered their time, helping us to maintain the quality of the SEAC reports. And finally we would like to thank the authors who have worked so patiently with us in the preparation of these *Proceedings*.

The editors
Stockholm, May 2003

The virtue of looking in another direction

Bengt Gustafsson*

The Danish expedition to Arabia left Marseilles in early June 1761, on board His Majesty's ship *Grönland*, the Greenland. The expedition was sent out by the Danish king to explore Arabia, in its fairy-tale shimmer *lykkelige Arabien*, happy Arabia, and leader of the expedition was a Swede, Peter Forsskål, a well-known and controversial pupil of Linneus at Uppsala. After a few days' journey from Marseilles, four English ships were sighted. Since England was at war with France, a fight was probable. The ship was made ready for battle. The cannons were put in position, handguns were distributed among the crewmembers. The commander of the English flotilla signalled that he demanded to hail the three merchants escorted by the *Grönland*. The demand was rejected by the commander of *Grönland*, and the ship was ordered ready.

Then a grand event occurred. It was June 6 in 1761, and the planet Venus was to pass across the solar disk. This only happens a couple of times each century. In addition to being rare, the phenomenon was also important because it offered the possibility of accurately measuring the distance scale within the solar system. Thus the astronomers in Europe had agreed on measuring the passage from as many different places as possible. Of particular significance was the determination of the times when the planet entered the solar disk and when it left it. The passage across the sun takes a few hours.

Now just before the fight is about to start, the geographer and land-surveyor of the expedition, Carsten Niebuhr, puts up his instruments—an astrolabe and a telescope—on deck, among the sailors who are busy preparing the ship for the battle. Unfortunately, he complains in his diary, the shakings and tremblings of the ship, in spite of the calm winds, prevented him from performing the measurements with the accuracy desired.

The Danish author Torkild Hansen, who has written about this episode in his fascinating book on the expedition, *Det lykkelige Arabien*, has the following comment:

> There is something touching in this scene of the serious astronomer standing on the foredeck, quite engaged in his instruments, while the sailors worry round him, and the English battleships lie waiting some cables away on the blank sea. One reason why the world has not yet gone down is perhaps that, also in the most dramatic moments, there is almost always one or two that continuously look in another direction. On some circles of sand, on a gable of a house in Delft, on board the ship where cannons are made ready for delivering their arguments on life and death, there is a man who is totally absorbed by the observation of a Venus passage.

At the very same moment on this early morning, there were festal spirits here at the Stockholm Observatory,[1] Many had found the way since the secretary and astronomer of the Royal Academy, Pehr Wilhelm Wargentin, had announced the Venus passage in the newspaper *Inrikes tidningar*. Here were the queen and the young crown prince Gustav, here were ministers and foreign ambassadors, and numerous male and female spectators. The murmurings made it difficult for the men at the telescopes to hear the time clicks of the clocks. One of them was Wargentin himself, another was the Uppsala professor Samuel Klingenstjerna.

The Royal Academy of Science in Stockholm was through Wargentin tied into a network, the large international project to determine the distance to the sun by studying this Venus passage and the following one some years later—these rare events show up in pairs. The Royal Society had sent expeditions to St. Helena and the Cape of Good Hope, the Academy de Science to India and Tobolsk. Wargentin had also organised observations in other places: the Finnish Lapland, the astronomers at

[1] This and similar references are to the beautiful Observatory Museum, in which the conference was held. Popularly known in Stockholm as the Old Observatory (Gamla Observatoriet), it was built by the Royal Swedish Academy of Sciences. Completed in 1753, its 250th anniversary will be celebrated this year, 2003 (ed. note).

Lund and Uppsala observed the passage, and amateur astronomers all over Sweden made their contributions in good order. In Torneå, in the far north, the remarkable enthusiast and local district judge, Anders Hellant, investigated the phenomenon together with a sheriff, a foundry proprietor, an alderman and, according to the sources, "all local nymphs". Among those were probably Hellant's housekeeper, mistress, and assistant Mrs. Britta Widte, who must be regarded as the first Swedish female astronomer.

From all available observations, Anders Planman, who became professor in physics in Turku in Finland a few years afterwards, could determine the distance to the sun, which agrees with the present determinations within 10%.

For a later colleague, there is much to contemplate in this episode from the first great epoch of Swedish astronomy. One striking circumstance is the ambition to anchor the activities in international collaboration. In this work to systematically make fruitful international contacts, Per Wargentin, the first director at this Observatory, was a master. "We all have one and the same sky and thus the same business," he wrote. It was not only the importance of participating in large-scale international projects, but also of asking for proposals for new projects and suggestions concerning on-going work. His teacher Anders Celsius, astronomer at Uppsala and inventor of the temperature scale, was the pioneer who maintained an international standard for his students and successors in this respect, through his travels and contacts.

The second circumstance that strikes a much later colleague is the significance of simple goals for large common projects. This need to determine the distance to the sun, on which the entire astronomical distance scale is founded, was an evident task to agree and unite about. With this you could make your colleagues go sleighing through the wilderness of Lapland with their instruments; you could get the aristocracy and other upper class people in Stockholm to walk up the hill to the Observatory in the early morning and, what was more important, spend considerable sums to support the project; you could get diligent amateurs to build their own instruments, like the priest from Småland, Sten Wikbom in Näshult, and Stenbert, who later tried to bribe Wargentin with a big cheese in order to persuade him to let the Academy print a description of his homemade quadrant. Yes, the simple goal to determine the distance to the sun got the land surveyor Carsten Niebuhr to risk his life and provoke his fellow travellers on board the *Grönland* in the Mediterranean on June 6, 1761.

One may still ask whether Niebuhr's conduct, and whether Torkild Hansen's appreciation of it and of the virtue of "continuously looking in another direction", would have been accepted in Stockholm, as the Royal Academy of Science was really directed towards making direct and practical use and to service in society, more than any other Academy. According to the first paragraph in the statutes of the Academy "all sciences and arts that are of some real use in public affairs should become subjects of its attentive and affectionate endeavour". This was stated repeatedly, for instance in the speeches given by the president or chairman of the Academy when he left his post. This happened often, since the election period for the president was only three months. For example, the captain in the Royal Engineers, Sven Ljungquist, describes in his speech on such an occasion in 1752, how scientific research lives in a compact with practical life and how the experiences of farmers and craftsmen enrich science: "a weaver at his loom, a fuller in his fullery, a presser at his press, a brewer at his vat, may all make useful experiments". But conversely, also practical activities must get ideas and methods from the natural sciences.

Even if housekeeping in a wide sense and the demands of economical life had the highest priority, the Academy did not neglect pure science. Fundamental research and its applications formed an entity that could not be split.

For Sweden and the Swedes, who had lost an empire a few decades before, the interest in practical matters and the use of science was patriotic, and this form of practical patriotism leavened all the daily work of the Academy. Work was carried out in farming, improving agricultural implements and manuring, rooting out of moles and bee-farming, silkworm breeding and cultivating colorific plants, forest economy and iron manufacturing, fireproof clays and roofing materials, organ building, fire extinction and air conditioning, public health and emigration issues, and uncountable other useful things.

The astronomers of the Academy were not unaffected by these ambitions. The first secretary of the Academy, Pehr Elvius, was the nephew and the son of an astronomer and first cousin of Anders Celsius. Elvius was an astronomer, but he also became engineer and mechanic. His great interest was water mills and mining mills. His successor as secretary was Pehr Wargentin who, aside from being an excellent astronomer and director of this Observatory, was a pioneer in demography. Words like *nativity* and *mortality* are due to him. He was in turn succeeded by Henric Nicander from Uppsala, who spent quite some time in constructing ploughs.

Also the astronomers from Uppsala and Lund who were members of the Academy took part in its practical programme. Mårten Strömer wrote his first contribution to the publications of the Academy on a method to prevent trees from being damaged by the frost in cold winters. When he finished his service as president in 1756, he spoke 'On the connection between astronomy and the art of navigation', and he later led the surveying for nautical charts in Swedish waters, a difficult task due to the rich archipelagos. Anders Celsius contributed, like several of his colleagues, to geodesy and cartography and made early contributions in other areas, not the least being meteorology, even if the real pioneer in this field was Erik Burman, who was also an Uppsala astronomer. Already in the 1720's, he established a network of weather stations with thermometers and barometers, which he also tried to spread internationally. These endeavours were continued at Uppsala and at other places by Celsius's successors and the astronomers in the Academy like Bengt Ferner, with Celsius's aim to "from an extensive set of observations be able to predict the weather as accurately as the prediction of solar eclipses". There is still some way to go in this. Bengt Ferner also constructed new ploughs and new types of harness; and he participated actively in the development of air conditioning or 'weather-exchange machinery' (väderväxlingsmaskiner) for battleships, which was an important undertaking, since a great fraction of the sailors who died in the naval battles of those days died from poisoning from exhaust fumes on the battery decks.

Many more examples may be given, but the conclusion should be clear already—Swedish astronomers during the 18th century took very active part in the use of their knowledge and skills "in the noble intention to apply them for public benefit", as a leading politician of the time wrote (Carl Gustaf Tessin). All this useful activity must, from time to time and in the everyday life of the scientists, have come into conflict with the ambitions to contribute to fundamental science, both individually and for the Academy as a collective. But the impression is still that the enthusiasm of the scientists to contribute useful applications for the housekeeping of the kingdom was generally great, so great that it even stained the language of Wargentin: he could write, for instance, in a letter to Pehr Elvius about his very significant studies of the satellites of Jupiter as "studies of the secrets of the housekeeping of Jupiter".

The members of the Academy showed in action that "it did not wish to abstain from any efforts by which she could serve the public", as one may read in a protocol from 1763. This was true not least for the astronomers. Still this is not a complete and satisfactory explanation for the remarkable fact that the Royal Academy of Science early on decided to establish an observatory as its first institution. One may think that a chamber of models for machinery construction, or an experimental station for agriculture would have been more suitable and fitting to the programme of the Academy. Why an observatory?

One may see several possible explanations. One was that two of the models and examples, the French and the Prussian academies of science, had founded observatories. On the other hand, neither of these institutions had such a utilitarian profile as the Swedish Academy. Another reason is, of course, the impressive contemporary success of astronomy, especially through Newtonian mechanics, which also in an obvious way connected astronomical phenomena to the machines of technological progress. Astronomy showed the way towards a rational and useful understanding of Nature. Moreover, astronomy had existential significance. In the natural theology of those days, the works of the Creator, from the large-scale structures in the solar system to its most minute details like the eyes of insects, were illustrations and expressions of his eternal wisdom and his loving consideration. It was written as an inscription in the observation hall in this building that the Observatory was built to "from the heavenly orbits find inspiration to our reverence and awe for the Lord of the Heaven and the Earth". These words were not empty. Many Academy members, and not least Linneus, would certainly have agreed with them, although certain sceptics like Anders Celsius might have had a different opinion.

Still, in retrospect, it is rather remarkable that the Royal Academy of Science succeeded in getting this Observatory financed by state money and erected so rapidly, not least because Anders Celsius had just got the University Observatory at Uppsala finished and furnished with instruments. Celsius argued skilfully for his institution in terms of its usefulness for the country: material for the Swedish general geographical atlas would be collected there, the calendars would be improved, meteorological journals be kept, teachers of navigation taught, etc. Similar arguments must have been used by Elvius and other influential Academy members in advocating the new Observatory in Stockholm. A guess is that Stockholm's Observatory could be founded as a result of a strong and well-coordinated lobbying activity from a number of influential individuals who launched astronomy as a very significant area of activity, in spite of the fact that it "continuously seems to look in another direction" as seen in our perspective.

I mentioned that the Venus passages occur about twice per century and pair-wise. On June 3, 1769, there was a second one. Also this time observations were made through large international campaigns. A great number of expeditions were sent abroad. One of them was an expedition sent to Tahiti, in the Pacific, from the British royal society and admiralty. The name of the ship this time was the *Endeavour*, and a leading scientist onboard was Daniel Solander, another of Linneus' students. The captain of the ship was the young naval lieutenant James Cook. When on Tahiti there were problems, not the least of which was the seizure of the most important instruments by the natives of the island. After considerable negotiations Cook and Solander could buy back the instruments, but they were now in pieces, and it was not an altogether simple task to put them back together. On the long way back to Britain, Cook made an extra tour around the southeastern part of the ocean. Finally back in London after three years, Cook was strongly criticised for not having fulfilled the goal of this endeavour—to acquire good observations of the Venus passage. On the other hand, he was soon graciously forgiven, since his expedition had some unexpected spin-off effects, as we would call them: the mapping of the east coast of Australia and the rediscovery and mapping of New Zealand and several smaller groups of islands, one of them also named after Solander. From this we learn that the usefulness of scientific endeavours are often very difficult to judge beforehand—the fundamental use of science is that new things are tried systematically, which may lead to quite unexpected results. We do not know what happens when we look in other directions.

I shall now move about 100 years forwards in time. In 1871, a new astronomer and observatory director arrived here, Hugo Gyldén from Finland. He had a background from the great Pulkova Observatory outside St. Petersburg and was early regarded a *wunderkind.* When he came here, he was 30 years old and was internationally known as an excellent theoretician within celestial mechanics, with work on the famous three-body problem and pioneering results in galactic dynamics, numerical orbit calculations and, indeed, chaos research, as we now call it. There was, however, one more characteristic—Gyldén was a radical liberal with great interests in social questions. These interests brought him to service for the state—he became a member of committees working on a system of pensions for everyone, for collecting general statistics on employment and accidents at work, and for arranging a system of insurance for workers.

Gyldén visited a doctor and physiotherapist, Gabriel Branting, who lived here near the Observatory. His son, Hjalmar Branting, showed early interest in astronomy and at 14 he was allowed to assist in the observations until the early mornings. "Branting appreciated the aristocratic and rational world view of Gyldén, quite unaffected by traditional superstition", as one of Branting's biographers puts it. Later in his life Branting himself characterised astronomy as "a heretical science, more so than all others". Letters show that Gyldén indeed was a fatherly friend and mentor to Hjalmar; moreover, he clearly showed his appreciation of Hjalmar as a devoted, talented, and soon very skilful observing astronomer. And Hjalmar finally expressed his thanks to his mentor in an obituary in 1896.

Hjalmar went to Uppsala to study astronomy, but there the teaching was stopped since the professorial chair was temporarily vacant. After one year, he made a journey abroad for purposes of study, a trip laid out by Gyldén to Turku, Helsinki, Tartu, Pulkova, Berlin, Paris, Frankfurt, Jena, and Hamburg. The journey was, however, at least as important for his political development as for his astronomical contacts and development. After a few more turbulent and adventurous years at Uppsala, when little was read but much of his father's money was spent on parties and prostitutes, he left the university and wrote at the end of the spring semester in 1882 to a friend, Karl Staaf: "Yesterday I talked to Gyldén, and he had not the least objection against my plan to skip Uppsala and settle at the Observatory. Conversely, he thought it was very reasonable to go back to astronomy. So, it is agreed and decided. This autumn, maybe already in August, I shall take residence at the Observatory for, as I hope, a period of many years".

It was most probably good for Branting that this choice was made. After eight terms at Uppsala, he had not even a bachelor's degree. The spirits and the politics had diverted him. His inherited fortune was hazardously reduced and, moreover, he now suffered from a venereal disease, probably gonorrhoea. Through his older colleague at the Observatory, the amanuensis Erik Jäderin, Branting got in touch with Jäderin's sister Anna, and they later married, and made a nice couple. Branting then started a career as journalist and finished his daily and nightly work at the Observatory in 1884. Later he became the first chairman of the newly formed social democratic party and, in 1921, he became the first social democratic prime minister. He was certainly one of the most important founding fathers of modern Swedish society.

In his hero biography of Branting, Zeth Höglund writes:

> During the years 1882-84, Branting was amanuensis at Stockholm Observatory and thus still planned a scientific career. But, in the end, the political struggle attracted him more, his inner calling became irresistible. There is a famous passage by the Greek philosopher Anaximenes, in a letter to Pythagoras, which appears in my mind: "How could I take interest in the secrets of the stars, when I constantly see death and slavery? " How should Branting be able to spend his life in exploring the heavens, however beautiful this occupation may be, when the earth carried so much unhappiness because society was so unfairly established? Was it not his duty primarily to fight against all the spiritual destitution and social slavery that disgraced his fatherland? He made his choice and he never had to regret it, because he chose the greater calling.

Even if one must question this idealistic description of Branting's choice to leave the Observatory, his view was obviously very different from that which Torkild Hansen praised and saw in Carsten Niebuhr's observing of the Venus passage on the Mediterranean. Both attitudes must be respected. Both attitudes have survived and are clearly present among us in the scientific community and among politicians and numerous other citizens. One of them may be seen as an elitist view, an appreciation of science per se, on one hand; at best the idea that long-term solutions can only be obtained from fundamentally new approaches, which can not be planned today. The other perspective is that the social problems and needs are so urgent that delaying them for solving more esoteric problems is not acceptable. But what can then these different histories and personal attitudes tell us today?

I would strongly suggest that we do not pull out the two different attitudes to the extremes. We can sooner regard them as complementary. Perhaps the heroic, romantic 19th century view of science and scientists has widened the gap between these two perspectives in a way that is, in fact, not even historically correct. Thus I have been told that Hjalmar Branting actively participated in the Swedish Astronomical Society here in Stockholm. And Carsten Niebuhr did not always look upon the sky. Peter Forsskål, the leader of the expedition to Happy Arabia, which turned out not at all to be so happy, died, as several other members of the expedition did, from malaria and hardship. Carsten Niebuhr, our astronomer, took on the task to lead the expedition with care, wisdom, and a sense of reality through the tough fatigue, back to Denmark. One should look in other directions, or at least some should do it, but we do not always have to do so.

*Department of Astronomy and Space Physics, Uppsala University, Box 515, SE-751 20 Uppsala, bengt.gustafsson@astro.uu.se.

The erratic ways of studying astronomy in culture

Stanislaw Iwaniszewski*

Abstract
In order to prove that astronomical objects and phenomena cannot be treated as independent categories in archaeoastronomical research, the use of the terms 'cosmology', 'world view', 'cosmovision', 'outlook', and 'understanding' are discussed and deconstructed.

Discussion
Archaeoastronomy has been traditionally defined as an interdisciplinary study bridging the sciences and humanities, the object of which is to investigate the relationships between man and the sky. As such it is a mixture of different particular approaches reflecting a diversity of standpoints of the scholars involved, which range through the sciences to the social sciences, arts, and humanities. Such diversity is seen as a very positive element, enabling continuous exchange of different opinions. It also involves a range of conflicting perspectives because the sciences and humanities rely upon different research paradigms.

For a long time I have believed that such diversity requires the invention of some all-embracing theory. I think I was very naive in thinking that such a thing was ever possible. Instead of a single theory of the cultural meaning of astronomy, we must look for several more modest alternatives. As long as archaeoastronomy is considered to be interdisciplinary, it will remain a convenient playground for a diversity of topics rather, and not theories. Its interdisciplinary character makes any deep discussion impossible. Let me then pose my first question in the following way: *Is it the interdisciplinary character of archaeoastronomy that impedes any deep and follow-through discussion leading to the establishment of its theoretical foundations?*

Interdisciplinarity means that any single or particular phenomenon may be studied from the standpoints of different disciplines. As each of them uses its own methodology and epistemology, some kind of a common language has to be established in order to assure effective communication within the scientific community. In our case this does not refer to the definition of what a particular astronomical object or phenomenon is, but rather to the nature of the social process attached to it.

I will illustrate this by referring to the uncritical and irresponsible use of terms such as 'cosmology', 'world view', 'cosmovision', 'outlook', and 'understanding'. For most scholars these concepts remain unproblematic. Cosmological beliefs, world-view patterns, and outlook paradigms are treated as synonymous, and I suspect that scholars inspired by the possibilities of examining material cultural items as source material for the study of past views of the world might find those concepts too diffuse to be of use in cultural astronomy. When a more formal definition of world view is required, it usually is formulated according to functionalist, structuralist, and/or phenomenological perspectives. This does not necessarily imply that past societies viewed themselves in the same terms.

Some British and American social scientists use the term 'cosmology' to refer to the human concepts of the universe (Mendelson 1976; Cordeu 1984: 190). In this they are followed by astronomers, for whom 'cosmology' is the study of the origins and nature of the universe or a disciplinary field in which theories of the universe are hypothesized, tested, and developed. Scientifically motivated scholars prefer this term, possibly because its meaning is very close to the use of 'cosmology' in the sciences (Jones 1972: 87). Ancient 'cosmologies' are considered as rough, pre-scientific, or non-scientific equivalents of modern 'cosmologies' and viewed as representing cognitive and functional/operational concepts of ancient cultures (Fritz 1978), in more or less the same way as physics, astronomy, and philosophy of science define 'cosmology' today. By this they metaphorically express their aspiration of improving methodology. Dissatisfied with previous epistemologies, social scientists tend to borrow methods from the natural sciences and to equate the discipline's improved methodology with the adoption of scientific methodology.

Having established equivalents between native conceptual frameworks, anthropologists started to search for the rules by which societies constituted their models of the world (Colby 1975), assuming that the universal perception of natural events, processes, or phenomena unifies individuals' different representations within the bounds of a particular society. Such universal cosmologies are thought to be based on objective and unchangeable features of material and natural environments rather than on specific actions of particular social agents. Methodologies based on universally recognized generalizations encourage cross-cultural comparisons and neglect the influence of social conditions on shaping cosmologies (Colby et al. 1981). 'Cosmology' becomes like a type of impersonal dimension of social systems that either already exists in human minds or stems 'naturally' from the relationship between society and the natural world around it. Societies interpret nature and material culture as containers upon which significance is to be inscribed. This nature-culture dichotomy reveals the western conceptualization of knowledge, which equates cognition with scientific knowledge, privileges scientific discourse over other types of discourses, and excludes other modes of cognition and knowledge from the domain of science. Society is interpreted as a coherent system in which individuals come up with similar statements regarding the world and the position of man within it. 'Cosmology' is not socially problematic and reduced to narrow cognitive issues. It appears to be advantageous for those scholars who consider that functional and cognitive aspects of cosmologies are of particular interest because they articulate the behavior of the people who make use of them (Reichel-Dolmatoff 1976). Astronomical alignments are best explained in terms of their use as calendrical markers associated with specific agricultural, ceremonial, or other important social cycles.

Cultural anthropologists often use the term 'world view' in the sense of the culturally constructed knowledge of the universe (Kearney 1975; but compare Redfield 1952; Foster 1966; Dundes 1971; Cordeu 1983). This approach emphasizes the social (visual, Ong 1969) construction of reality, defines the social constitution of knowledge, and avoids the strictly functional and cognitive approach described above (Cordeu 1983: 286). Again, however, world-view models are explained in terms of shared-knowledge systems, assuming that particular members of social groups or communities are the individuals who know how to act in various contexts. This approach leads to the conceptualization of shared world views in terms of social or ethnic identities and may address important questions in terms of the exercise of power relations through the establishment of dominant world-view models. World-view models are based on culturally constructed representations of the universe (especially in the concept of *Weltanschauung*). Such a position is neglected by a number of astronomically trained archaeoastronomers because it involves a presupposition of the non-objective nature of systems of knowledge. Some anthropologists emphasize the passive role of an individual, as Malinowski puts it (1922: 517): "every human culture gives its members a definite vision of the world". Others imply man's more reflexive and active attitude, for example Redfield (1953: 85): "man is aware of the structure of things". Still others emphasize the passiveness of man's environment instead of man as a passive individual (Galdston 1972: 95).

Latin American scholars prefer to use the term 'cosmovision', which is a somewhat imprecise translation of the English term 'world view'. Broda (1982: 81) introduces notions of structure, systems, and totality, defining cosmovision as "the structured view in which the notions of cosmology are arranged into a systematic whole". This approach accounts for the presence of political, social, and economic dimensions in the constitution of cosmovisions. The fact that astronomical objects and phenomena are universally perceived is exploited by the ruling class, which imposes its own cosmovision upon the rest of society; and those ideologized models of the universe provide the ruling class with the appearance of manipulating natural events and processes in such a way that it legitimizes and justifies the existing social order. As Broda argues, both the individual and the social group are active in shaping cosmovisions.

The non-objective and ideological manipulation of cultural and social reality is found among those scholars from the former socialist countries who use the term 'outlook'. It refers to rational and irrational components, underlines an evolutionary character of cultural ways of knowing, as well as includes the dynamics of social relations in the practice of its development. This approach accounts for individual versus collective outlooks and offers an evolutionary perspective starting with the most irrational and ending with the most rational models of the universe. Cognitive and social knowledge is conceptualized as resulting from material and historical circumstances.

Recently Jacques Galinier (1999) rejected the 'petrified' term of 'world view' and proposed the use of the concept 'understanding' instead. Emphasizing "the operations related to a 'symbolic logic' that proposes truth judgments based on the absence of a principle of non-contradiction and on 'synapsis effect", his concept of understanding brings together cosmovision, world view, and ethics. Although Galinier's approach involves an active individual and does not conceive of 'understanding' within the dualist framework, his stress on symbolical-logical operations privileges cognitive rules of structuring people's own being-in-the-world.

We can see, then, that our uses of terms such as 'cosmology', 'world view', 'cosmovision', 'outlook', 'understanding' denote different epistemological and ideological positions. Not only do they refer to different areas of human experience (together with the concepts of 'ethos', 'normative systems', 'ideology' they belong to symbolic parts of cultures), but they also denote conflicting methodologies (cf. discussions in Geertz 1968; Leichtman 1969; and especially in Jones 1972). My second question therefore is: *Is it the predominance of the utilitarian and cognitive model of astronomy (derived from the research paradigm of the sciences) that impedes any dialogue with humanists in the interdisciplinary framework of archaeoastronomy?*

In most studies, astronomical objects and phenomena receive an independent ontological status and are understood as categories separated from the human mind. As objects they have to be perceived, and as phenomena they must be understood. Observed, analyzed, measured, and quantified according to standards of Western science, they are conceptualized in utilitarian and functional terms. Ready to be discovered and appropriated by man, they remain separated from the reality of social practice and human culture. This orientation reflects the ideas of modern astronomy. What is being perceived and experienced in the present is based on the cognitive and functional framework that is entirely the product of generations of astronomers on the one hand and the result of the westernization of the political and economic relations in societies on the other. This Cartesian dualism, separating nature from culture and astronomy from social life, cannot become an epistemological premise on which archaeoastronomical methodology and theory are built. Although astronomical objects and phenomena belong to an independent physical space, they come into existence in a social sense through practice. In other words, it is a fact that a particular sunrise over a prominent peak A is observed on date B at site C; calculations back in time can prove whether this could have happened a few thousands years ago; and estimations can give us the idea of the precision of the observations involved. However this concept provides us with a framework for studying astronomy as an empty category. Sunrise observations are made by human beings and not by objective, omnipresent, and omniscient observers. In our interpretation we are dealing not with astronomical events belonging solely to a category of *nature*, but with astronomy as a social category as well. Social phenomena are research objects that are quite different from physical astronomical phenomena and, as many scholars have demonstrated (Giddens 1984), each of the two parts of research requires its own epistemology and methodology. From the viewpoint of social theory, astronomical objects and phenomena are in themselves meaningless until they are situated within a society. Although astronomical events are universally observed by human beings, they are not similarly perceived (i.e. explained and interpreted). By addressing social practices and social existence in relation to astronomical objects and phenomena, they become 'domesticated', loosing their 'objective', absolute, and unproblematic character. Particular sunrise observations become important not because they are observed from a given point on a certain date, but because they occur in human societies that through the dynamic relations between their institutional agencies, daily routine practices, and individual interrelations create fields of discourse from which those observations become significant (Barrett 1987/8). Astronomical alignments encoded in architectural monuments are situated within these fields of discourse, constantly produced and reproduced, interpreted and reinterpreted, and modified through social practice.

Furthermore, we cannot apply the same criteria to discover the meanings transmitted into our present-day practice. With great probability our theories regarding past peoples in terms of knowing subjects who act as if they knew our concepts of calculated risk, maximum efficiency and, last but not least, of equinoctial points are wrong. We must acknowledge that astronomical alignments and motifs in art objects were full of ideology, meaning, function, or utility for past societies.

Seeking criteria for proof has been a very important and yet unresolved problem in archaeoastronomy. How can we make presuppositions if we know that everything is relative? How can we take a decision as to what criteria of validation in social sciences were also valid in the past? The hopes of solution seem to me to be contained in a double structure of archaeoastronomy. While dealing with the astronomical, 'hard' component, archaeoastronomy can use the same criteria of falsification or validation as other exact sciences, i.e. hypothetic-deductive, or inductive approaches, and subject-object Cartesian dichotomy. However, in order to interpret astronomical facts in terms of social theory, this kind of scientific thought must be replaced with discourse of an hermeneutical character and without any strong dualism between subject and object. Accepting that archaeoastronomical research is a two-stage process, involving scientific paradigm first and humanistic paradigm later, I have tried to avoid fruitless discussions regarding its status. We should therefore not overlook the fact that, if we speak of a particular sunrise, we are interpreting astronomical phenomena already interpreted in terms of human existence and practice in the past. My third question is therefore: *Is any dialogue possible if astronomical objects and phenomena are never recognized as possible constitutive elements of human existence and practice?*

Conclusions

Archaeoastronomy consists of two conflicting epistemologies. Astronomical events and objects belong to the category of natural phenomena and demand their own epistemology and methodology. Social phenomena form a research area that is different in character from the physical processes of nature. Astronomy cannot be included in social theory as an independent category; rather it should be incorporated into social life with the help of social mediation, and this mediation is established through social practice. What is studied are social practices performed by different human groups in particular social contexts against the common background of astronomy.

* Division de Posgrado, Escuela Nacional de Antropologia e Historia, c. Periferico Sur y Zapote s/n, Deleg. Tlalpan, 14030 Mexico, D. F., siwanisz@yahoo.com.

References

Barrett, J. C. 1987/1988. Fields of discourse: Reconstructing a social archaeology. *Critique of Anthropology* **7**, 5-16.

Broda, J. 1982. Astronomy, cosmovision, and ideology in pre-Hispanic America. In *Ethnoastronomy and Archaeoastronomy in the American Tropics* (Annals of the New York Academy of Sciences 385), ed. A.F. Aveni and G. Urton, 81-110. New York.

Colby, B. N. 1975. Culture grammars. *Science* **187**, 913-919.

Colby, B. N., Fernandez, J. W., and Kronenfeld, D. B. 1981. Toward a convergence of cognitive and symbolic anthropology. *American Ethnologist* **8**, 422-450.

Cordeu, E. J. 1983. Los nexos cosmovisionales de la demencia. *Suplemento Antropológico* **18**, 285-304.

Cordeu, E. J. 1984. Categorías básicas, principios lógicos y redes simbólicas de la cosmovisión de los indios Ishír. *Journal of Latin American Lore* **10**, 189-275.

Dundes, A. 1971. Folk ideas as units of worldview. *Journal of American Folklore* **84**, 93-103.

Foster, G. M. 1966. World view in Tzintzuntzan: Re-examination of a concept. In *Summa Anthropologica, homenaje Roberto J. Weitlander*, 385-393. México.

Fritz, J. M. 1978. Paleopsychology today: Ideational systems and human adaptation in prehistory. In *Social archaeology. Beyond subsistence and dating*, ed. C. L. Redman, M. J. Berman, E. V. Curtin, W. T. Langhorne, Jr., N. M. Versaggi, and J. C. Wanser, 37-59. New York.

Galdston, I. 1972. Comments on "World views: Their nature and their function" by W. R. Jones. *Current Anthropology* **13**, 95-97.

Galinier, J. 1999. L'entendement mésoaméricain. Catégories et objects du monde. *L'homme* **151**, 101-122.

Geertz, C. 1968. Ethos, world-view and the analysis of sacred symbols. In *Every man his way. Readings in cultural anthropology*, ed. A. Dundes, 301-315. Englewood Cliffs.

Giddens, A. 1984. *The constitution of society.* Oxford.

Jones, W. T. 1972. World views: Their nature and their function. *Current Anthropology* **13**, 79-109.

Kearney, M. 1975. World view theory and study. *Annual Review of Anthropology* **4**, 247-270.

Leichtman, M. 1969. World views: Their nature and their role in culture. *Current Anthropology* **10**, 470-471.

Malinowski, B. 1922. *Argonauts of the western Pacific.* New York.

Mendelson, E. M. 1976. Concepción del mundo. In *Enciclopedia Internacional de las Ciencias Sociales* vol. 2, ed. D. L. Stiles, 690-692. Bilbao.

Ong, W. J. S. J. 1969. World as view and world as event. *American Anthropologist* **71**, 634-647.

Redfield, R. 1952. The primitive world view. *Proceedings of the American Philosophical Society* **96**, 30-36

Redfield, R. 1953. *The primitive world and its transformations*. Ithaca.

Reichel-Dolmatoff, G. 1976. Cosmology as ecological analysis: A view from the forest. *Man* **11**, 307-318.

Swedish calendar staffs

Sven-Göran Hallonquist*

The history and geographical distribution of runic staffs

In Europe the Roman calendar tradition had been preserved and developed within the church. This advanced time-keeping tradition was spread on parchment with Christian missionaries. Calendar staffs, made of wood and with text in runes (Fig. 1), are a simplified variant of this calendar tradition. They were most likely used exclusively by a small educated class, as an aid to memory, at a time when there were within the church considerably more comprehensive calendars written on parchment.

The forms of the runes on the staffs indicate that their origin can be dated to the 11thor early 12th century in the case of Linköping's diocese, which takes us back to the time for the establishment of this bishopric. In Skara diocese, which is older than the one in Linköping, having been most likely established in the 1020's, the calendar staffs are unlike those from the rest of Sweden in that they do not have runes as golden numbers (*gyllental,* see further below), but continental linear symbols built up on the basis of five lines or pentad line symbols (Fig. 2). This fact strengthens the dating of the origin of runic staffs to this time. The diocese of Västerås, which is later in date than that of Linköping, also has later runic forms.

In Denmark and Norway, which were converted to Christianity long before Sweden, there are also calendar staffs. These, however, do not have runes, except in a few late cases, which should possibly be seen as Swedish influence.

Runic staffs exist in many different shapes, e.g. different kinds of swords (most likely the oldest shape), different board-like forms, and also book-like shapes (boards sewn together like a book). The golden numbers and also the placement of some of the feast days on the staffs have their origin in the different diocesan calendars, but with the passage of time the golden numbers were often miscopied or the minor feast days were changed to meet local needs or those of later times (see further below). The combination of outer form and calendric content makes it often possible to place runic staffs in time and place.

In Sweden the use of runic staffs spread to the peasantry during the 16th century, most likely due to the efforts of the church. As a help in the education of the peasantry concerning calendric matters, considerable information on the subject was published in the Swedish psalm books of the 16th century.

In addition to purely calendric information, runic staffs often have cubit, foot, and inch markings and, often as well, the so-called St. Peter's game.

In the 17th century the use of runic staffs began to decline due to the yearly publication of the almanac. The latter are richer in content, easier to use, and also more reliable. Despite several attempts towards the end of the 17th century on the part of the state to revive peasant use of the staffs, it died out except in certain areas such as Upper Silja (Ovansiljan) in Dalecarlia, where it still existed in the 18th century. The staffs continued to be used by those Swedes who had long inhabited areas outside modern Swedish borders, i. e. the island of Hiiumaa just off the coast of Estonia, and also by those expelled to Ukraina (who used runic staffs until the 20th century).

At Uppsala University the educated elite were taught the construction and use of the runic staff from the end of the 17th century. Therefore there are many beautiful examples existing from this time, fine sword sheaths, miner's axes, and walking canes. Thus from this time on the runic staff became simply a beautiful status symbol for the educated.

At present the ca 800 authentic runic staffs preserved are being inventoried and documented by the Runic Department of the Central Office of National Anqituities. This work is expected to take eight to ten years, unless more scholars can be included in the project (especially in the case of the Uppsala material).

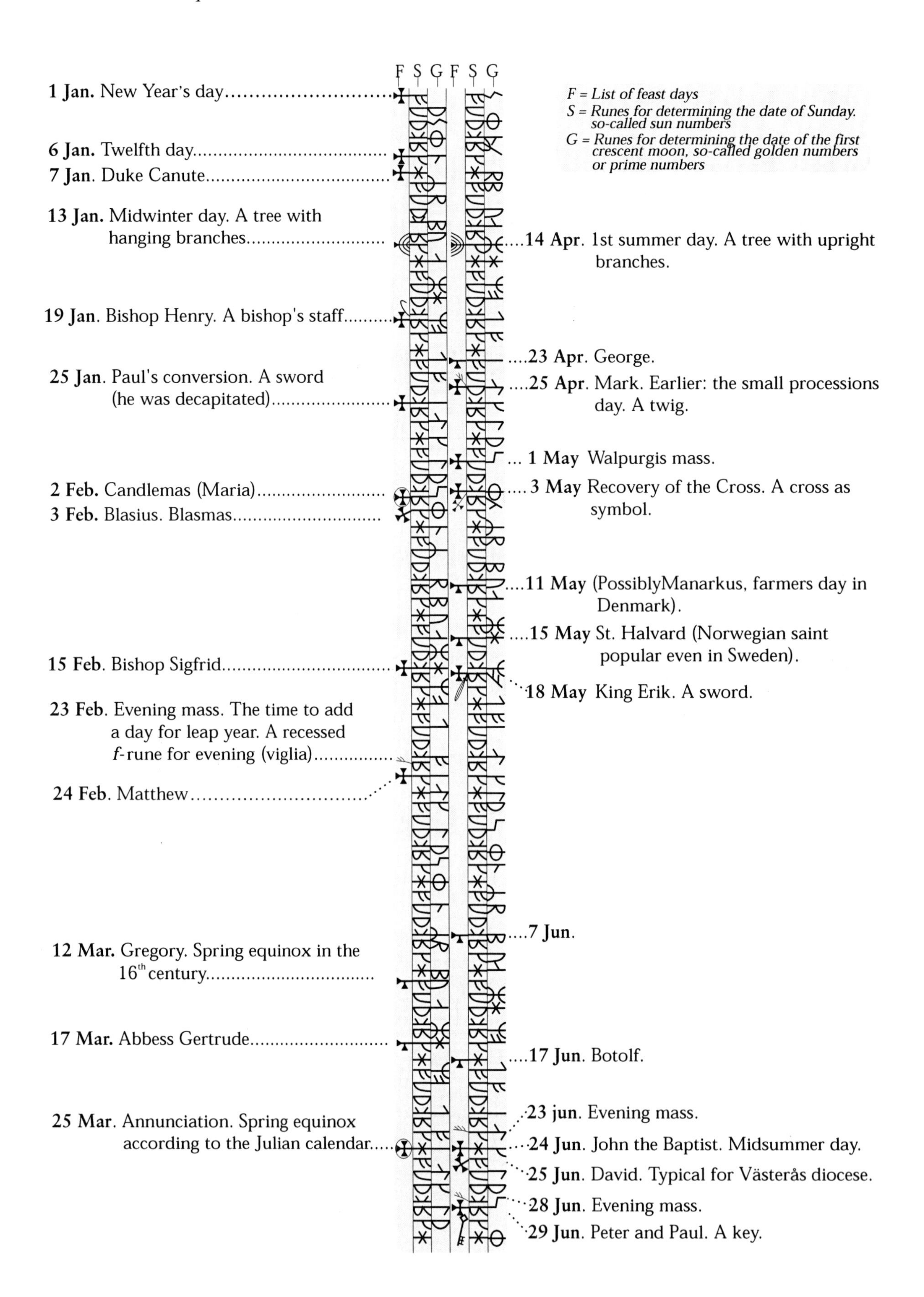

Fig.1. A runic staff typical of Mora (Dalecarlia) at the end of the 16th century. This side runs from 1 January to 1 July

The organisation of the runic staff

The letter *F* in the first row stands for *festdagslängd* (list of feast days). This row contains symbols denoting the feast days fixed by the church (diocese), which the peasants were to celebrate (Fig. 1). The symbols ⊕ ✢ ᚠ denote the days sacred to Maria, other feast days, and evenings. Often attached to certain of these feast days are the attributes of the saints, orthodox ones or variants originating with the peasantry, e.g. a key for St. Peter's day on the 29th of June.

The symbols ⊣ ᛏ , which denote minor feast days, are often the most instructive for understanding the yearly rhythm of the local peasantry, as they do not denote days set aside for celebration by the diocese, but rather days chosen at times more convenient for the peasantry, e.g. for the actual spring equinox, markets, or other events. Therefore the placement of these symbols frequently shifts in time and place. Often encountered on the runic staffs is an older, probably pre-Christian division of the year, frequently indicated by tree symbols of different kinds, e.g. trees with hanging and upright branches on the Mora example (Fig. 1).

Until the middle of the 17th century, all dates were referred to with respect to the feast days, e.g. 'two days before Gregory's' instead of 'the 10th of May', as we say today. On older runic staffs we also see small marks or lines in this row with refer to so-called unlucky days, when one should be especially careful.

The letter *S* in the second row stands for the possible dates for Sunday. The row contains the first seven runes of the runic alphabet (*futharken*), ᚠᚢᚦᚮᚱᚴᚼ, repeated 52 times (= 364 days), and they denote the dates on which Sundays could occur. The fact that runic staffs often show only 364 days instead of 365 reflects a pre-Christian so-called week-year. For the year 1317, the year of the Nyköping Banquet, the *u* rune (ᚢ) was the letter for Sunday, i.e. wherever this rune occurred it denoted a Sunday in that year. In order to know which rune should be read any given year, the so-called sun cycle was often given on the calendar staff's narrow side. The sun cycle characteristic for Västerås's diocese:

ᚠ ᚦ ᚱ ᚼ ᚢ ᚮ ᚴ
ᚼ ᚴ ᚱ ᚮ ᚢ ᚠ ᚼ ᚴ ᚮ ᚦ ᚢ ᚠ ᚴ ᚱ ᚮ ᚦ ᚠ ᚼ ᚴ ᚱ ᚦ ᚢ ᚠ ᚼ ᚱ ᚮ ᚦ ᚢ

reads from left to right and includes, for example, the period 1296-1323, i.e. a period of 28 years. In 1324, the cycle began again. Every fourth position contains two runes due to leap year. In these cases the rune above is read until the added day for leap year, the 23rd of February in Sweden. Then for the rest of the year the rune below is read. There were also different counting rules in different periods and places, which made possible the determination of which rune would be the Sunday letter for any given year.

The symbol *G* in the third row stands for the runes used to determine the date of the first crescent moon, the so-called golden numbers (*gyllental*). This row contains the 16 first signs of the runic alphabet and also three so-called connecting runes, or 19 runes in all, which comprise, for example, the period 1311–1328:

ᚠᚢᚦᚮᚱᚴᚼᚾᛁᛆᛋᛏᛒᛚᛘᛦᛮᛯᛰ

Then in 1329 one began again from the beginning.

These runes are placed at the dates on which there would be a new moon. For example, the new moon was expected to occur in 1317 on all dates where the rune for *h*, ᚼ, is found. However, towards the end of the Middle Ages, an awareness grew that the calendar showed increasing inaccuracy, amounting to one day per 300 years, with respect to the new moon. Towards the end of the 16th century, in order to compensate for this inaccuracy, so-called corrected golden numbers were introduced, which would show the actual new moon. For this reason we often find in Dalecarlia at this time the runes:

ᚦᚮᚱᚴᚼᚾᛁᛆᛋᛏᛒᛚᛘᛦᛮᛯᛰᚠᚢ

instead of those above. The change led to a correction of three days. Even in this case there were counting rules for determining the so-called golden numbers for certain years.

The different phases of the moon were considered to have great significance for human activities and well-being. Consequently all planting should be done during the waxing moon and all harvesting during the waning moon. A moon cycle at this time was called a *tungel,* and there were different designations depending on when it occurred, for example *Disa-tunglet*, which determined market-time in Uppsala.

A similar system was used in England, the so-called Clog-almanacs (see references below), where the Golden numbers had a similar form to those in Skara (Fig. 2).

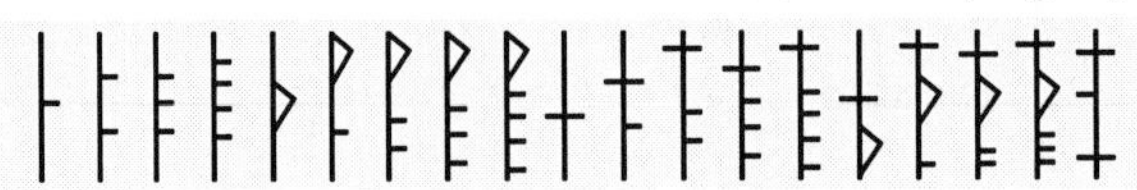

Fig. 2. Continental linear symbols

*The Royal Institute of Technology SE-100 44 Stockholm, Sweden,
goran.hallonquist@syd.kth.se .

References

Benneth, S. (ed.) 1994. Runmärkt: Från brev till klotter: Runorna under medeltiden. Stockholm.

Doxey, J. S. 1866–67. Notice of a Clog almanach from Wirksworth, Derbyshire. (The *Reliquary* **7**, 173 ff.

Harland, J. 1865. On clog almanacs; or runestocks, *The Reliquary* **6**.

Hasterup, K. 1985. *Culture and History in Medieval Iceland. An anthropological analysis of structure and change.* Oxford

Horne, J. S. 1948–1949. Staffordshire clog almanacs. *Transactions of the North Staffordshire Field Club* **83**, 13 ff.

Lithberg, N. 1953. Computus. Stockholm. *Nordiska Museets Handlingar* **29**

Magnusson, E. 1878. On a runic calendar found in Lapland. *Cambridge Antiquarian Society's Communications* **4:1**, 59–104.

Magnusson, E. 1877. The runic calendar. *The Academy* **12:291**, 515

Simpson, H.F.M. 1891–1892. On two rune prime-staves from Sweden and three wooden almanacs from Norway. *Proceedings of the Society of Antiquities of Scotland*, 358–378

Simpson, H.F.M. 1894–1895. Notes on a Swedish staff-calender, presented to the Museum by John Abarcrombie, dated 1710. *Proceedings of the Society of Antiquities of Scotland*, 234–240.

The pagan Great Midwinter Sacrifice and the 'royal' mounds at Old Uppsala

Göran Henriksson*

Abstract

According to the established interpretation the pagan Great Midwinter Sacrifice at Old Uppsala, in Uppland province, took place every ninth year. In our modern way of counting this means every eighth year. The starting date was determined by the full moon that occurred between 21 January and 19 February in the Julian calendar. After the introduction of Christianity the tradition with a great assembly at Old Uppsala at midwinter was continued by the *Disting*, which consisted of an assembly and a market. By combining historical data and calculations of the dates of the full moons within the Disting period, it has been possible to establish the exact years of the eight-year cycle. One such year was AD 852, the same year as St Ansgar's second missionary journey to Birka, the oldest Swedish town.

At the end of the 17th century, the farmers of Uppland were still using the so-called rule of King Aun, according to which the phases of the moon in the Julian calendar fell one day earlier after 304 years. Such displacements in the eight-year cycle took place in 1692, 1388, 1084, 780, and 476. The semi-legendary King Aun is considered to have reigned about AD 450-500 and to have been buried at Old Uppsala. The three 'royal' burial mounds there have been dated to AD 450-550. These mounds are oriented in such a way that they could have been used to regulate the sacrificial calendar.

Historical sources for our knowledge of the Great Midwinter Sacrifice at Old Uppsala

Vita Ansgarii, written before 876 by Rimbert (Rieper 1995), the successor of Ansgar as archbishop in Hamburg-Bremen, tells about the earliest known Christian Missions to Sweden, the two visits by Ansgar and some other priests to Birka in Lake Mälar 829-830 and 852. The ruler during the first visit (Rieper 1995: 37-40), King Björn, invited them because he wanted better relations with the Christian countries in northern Europe. One big problem in these relations was the Great Midwinter Sacrifice, including human sacrifice, which was celebrated regularly and was lead by the king. During Ansgar's second visit to Birka in 852 (Rieper 1995:66-76), he learned that there recently had been such a Great Sacrifice and that the Christian priests were no longer popular. The situation was similar in Denmark and Norway, but the Christian influence in Denmark was greater than in Sweden. *Chronicon,* written by Thietmar from Merseburg about 1000 (Trillmich 1957: 21 [Thietmar 1.17]), tells about the last Great Sacrifice at the ancient Danish heathen centre at Lejre.

Gesta Hammaburgensis Ecclesiae pontificium, by the German missionary Adam from Bremen, written about 1075, gives details about the Great Midwinter Sacrifice at Old Uppsala, the ancient main heathen cult centre in Sweden (Tschan 1959: 207-208 [Adam 4.26-27]). In the original text, Adam wrote "*Solet quoque post novem annos communis omnium Sueoniae provintiarum sollempnitas in Ubsola celebrari*". This means that every ninth year all the Swedish provinces had to send representatives to Old Uppsala for a common celebration. We must note, however, that when the early Swedes said *every ninth year* this corresponds to *every eighth year*, as they had no zero and counted the beginning of the first year as year one and reached year nine when only eight years had elapsed. It was the German Historian of Astronomy Otto Sigfried Reuter who first realized that *post novem annos* corresponds to *every eighth year* in our way of counting (Reuter 1934: 483-484). Most Swedish scholars, however, continue to believe that it was a true nine-year cycle. In fact this celebration took place every eighth year according to an eight-year cycle determined by the phases of the moon. The eight-year cycle is the shortest period after which the same lunar phase is repeated approximately on the same date, as eight tropical years = 2921.934 days and 99 synodic months = 2923.528 days. This means that the same phase of the moon will appear delayed by one and a half days after eight years. After 19 eight-year cycles (= 152 years), the cycle is shifted by a whole month, which was already mentioned by the Greek astronomer Geminos ca 70 BC (Aujac 1975: 47-58). He also explained why this cycle was called both the nine-year and the eight-year cycle in antiquity.

Aside from the above works, the earliest available written sources for early Swedish history are the old Nordic sagas and chronicles preserved on Iceland, written by Christian scholars such as Snorre Sturlason, who wrote a history of the kings of Norway ca 1230 (Monsen 1932). The first chapter gives the history of the earliest kings in Old Uppsala, from ca AD 150-200, as they were also the ancestors of the Norwegian kings. They were reckoned as offspring of the fertility god Freyr, who is said to be buried in Old Uppsala. An extensive collection of sources concerning early Scandinavian literature can be found in Sundqvist (2002: 39-62).

The Great Midwinter Sacrifice in Old Uppsala as described by Adam from Bremen

The place called Uppsala in the ancient texts is nowadays Old Uppsala (Gamla Uppsala), located a few kilometres north of the modern city of Uppsala. It was the main heathen cult centre in ancient Sweden, famous for its three large burial mounds (Fig. 1).

Fig.1. The 'royal' burial mounds at Old Uppsala from the south. Note the menhir on the flat *Tingshög* (thing mound or thingstead). From Rudbeck 1679b: plate 16, fig. 77.

It was a great triumph for the Christian Church that this terrible pagan place could be transferred to the Swedish archbishopric in 1164.

Some Christians had visited Uppsala and told Adam about the horrible Great Sacrifice, and he wrote that from every living creature nine male individuals were sacrificed to conciliate the gods. The bodies were hanged up in a sacred grove close to the temple. This grove was considered to be so sacred by the heathens that every tree was believed to have a divine force derived from the death and decay of the victims. For a period of nine days one man and seven male domestic animals were sacrificed. When the sacrifice was completed, one could see altogether 72 bodies hanging in the holy tree. Birds and rats ate the bodies when they disintegrated (Tschan 1959: 208 [Adam 4.27]). The excavator Bror Emil Hildebrand found some non-cremated human bones and bones from seven different domestic animals: horses, oxen or cows, pigs, rams, dogs, cats, and cocks in the filling material of the eastern mound. This may indicate that the holy tree was situated in the vicinity of the eastern mound (Lindqvist 1936: 206).

The importance of the Disting and the precise definitions as to when it should take place

The original meaning of the Disting was threefold; there should be: a great sacrifice for peace and victory for the king, a general meeting with representatives from all the Swedish provinces, and a major market (Granlund 1958: cols 112-115). At the general meeting important common political decisions were taken, such as election of a new king or solution of judicial questions that not could be solved at local courts. The participation of the representatives was compulsory, and Christian representatives who refused to come because of the human sacrifice had to pay a great fine.

The dates for the Disting were linked to the phases of the moon according to an ancient rule preserved in medieval texts. Already Tacitus had pointed out that important meetings among the

Germanic peoples must take place at the new or full moon (Hutton 1970: 149 [*Germania* 11]). In his *Historia de gentibus septentrionalibus*, written in 1555 during his exile in Rome, Olaus Magnus, the last Roman Catholic archbishop in Sweden, explained that the Disting was started at the full moon because the light from the moon facilitated travel to Uppsala during the short days at midwinter (Foote 1996: 203 [Magnus 4.6]).

The exact rule for determining the starting date of the Disting was given by Olof Rudbeck (1679: 68), professor in medicine at the university of Uppsala and a scholar with broad scientific interests: The moon that shines in the sky on Twelfth Day (6/1) is the Christmas moon and after this follows the Disting's moon. This means that the earliest date for the beginning of the Disting was 21 January (7/1+14 days) and the latest date was 19 February (7/1+29 days). The Disting started on the day of the full moon between 21/1 and 19/2, according to the Julian calendar. The corresponding interval for the beginning of the Disting in our modern calendar is 28 January-26 February. It may seem strange that this originally heathen rule was related to Twelfth Day, or the Epiphany, as in the rule for the start of the Disting in Magnus (Foote 1996: 203 [Magnus 4.6]). The explanation is that the rule for the dates of the Disting was related to the Christian calendar in the 12th century. At that time there was a shift by seven days between the Julian calendar and our Gregorian calendar that is closely related to the solstices and equinoxes. This fact also explains why the Swedish tradition says that the night of St. Lucia, 13 December, is the longest and darkest night of the year. If seven days are added to this date, we get the date of the winter solstice at that time. This fact indicates that the pre-historic Swedish calendar was closely related to the solstices and equinoxes and supports the results found in my earlier archaeoastronomical investigations of ancient monuments in Sweden (Henriksson 1983, 1989a and b, 1992, 1994, 1995, 1999 and 2002).

People of today may find it difficult to understand why such an exact rule was needed to determine the day for the sacrifice to the gods, but there seems to have been a long tradition in Europe that the gods must be worshipped on specific days. Geminos wrote that when the years are reckoned exactly according to the sun, and the months and the days according to the moon, then the Greeks think that they sacrifice according to the costume of their fathers; that is, the same sacrifices to the gods are made at the same time of the year (Aujac 1975: 50 [Geminos 8.15]).

An extensive discussion about the introduction of the eight-year cycle in Greece can be found in Ginzel (1911: 365-385).

The exact years in the eight-year cycle for the Great Midwinter Sacrifice

The oldest historical records give three conditions for identifying correctly the sacrificial years of the eight-year cycle:

1) There was a Great Sacrifice 851-853, just before the second visit of Ansgar at Birka, according to the Chronicle by Rimbert, written before 876 (Rieper 1995: 66-76).
2) The Great Sacrifices took place about the time of the vernal equinox before 1075, when Adam wrote in his Chronicle: "Hoc sacrificium fit circa aequinoctium vernale", which means this sacrifice took place about the time of the vernal equinox, i.e. at the end of February and beginning of March (Tschan 1959: 208 note b).
3) The Great Sacrifices at Lejre, the Danish counterpart to Uppsala, took place in January before 934, when they were forbidden, according to Thietmar of Merseburg, writing about 1000.

To solve this problem I computed all full moons between 28 January and 26 February in the Gregorian calendar for the period AD 200-1200. The only possible solution is an eight-year cycle including the year AD 852 as the year for the second visit of Ansgar in Birka. This means that the exact year for all the Great Midwinter Sacrifices can be computed as multiples of eight years counted from year 852 (Henriksson 1992, 1995).

Early Swedish history and the eight-year cycle

Not many exact years are known in early Swedish history, but those that we have correlate very well with the eight-year cycle (Table 1). This is easy to understand as all the important decisions were taken at the general assembly that took place every eighth year. It is also interesting to note that the tradition of having general meetings every eighth year continued after the last official Great Midwinter Sacrifice in 1084.

The historian Tore Nyberg (2000: 120-137), at the University of Odense in Denmark, has used this eight-year cycle to establish a chronology for early medieval Scandinavian history. He notes that the relations between the church and the people in Denmark and Sweden deteriorated close to the years in the eight-year cycle in which there was a Great Sacrifice and that all the important clerical meetings took place in middle of the interval between such years. He believes that there was a heathen political party that tried to continue the series of sacrifices as late as the middle of the 12th century. He also suggests that the murder of King Sverker the older, in 1156, was performed by members of this heathen party because Sverker may have tried to stop the sacrifice that year.

Year	Event
852	Second visit of St. Ansgar at Birka.
980	Foundation of Sigtuna, the royal city succeeding Birka, according to dendrochronological dating.
1060	The first bishop in Lund, Scania.
1076	King Anund had to abdicate because he refused to lead the Great Sacrifice at Old Uppsala.
1084	The last Great Midwinter Sacrifice in Uppsala. The heathen temple was burnt down.
1124	The bishop at Old Uppsala had to leave the country after a few months.
1156	King Sverker was murdered. According to Nyberg (2000: 136-137) this may have been an act of revenge from the last heathens in Sweden.
1164	The first archbishop at Old Uppsala.
1188	The minting of coins starts again in Sweden.
1204	At a general meeting in Söderköping it was decided that new towns should be built in Sweden.

Table 1. Important events in early Swedish History

Determination of the phases of the moon one year in advance according to the farmers' rule

During the Disting market in 1689, Rudbeck talked to a 90-year-old farmer from Uppland who demonstrated his old runic calendar staff (*runstav*). The farmer taught him the following rule for the shifts of the phases of the moon in the same month of the next year: *The phases of the moon will be shifted either 12 days backwards or 20 days forwards in the same month with 30 days the next year* (Rudbeck 1689: 652). Rudbeck demonstrated that this rule worked perfectly. For a modern reader it is also clear that he made the computations without zero. According to modern arithmetic we subtract 11 or add 19 days to the actual date instead of 12 and 20, as in the old rule. This rule could have been especially useful in the determination of the day of the Great Sacrifice one year in advance. For instance, if it were full moon on 8 February, the year before the Great Sacrifice, the next Great Sacrifice should be shifted from the latest date, 26 February, to the earliest date, 28 January. This could be directly observed in Uppsala as the sun sets along the three burial mounds on 8 February. Rudbeck also learned how the farmers determined the dates of the full moon as shown in Fig. 2.

When Snorre Sturlason visited the province of Västergötland during the spring of 1219, he was told that the Disting had earlier taken place in the month called Göje, corresponding to February, but after Christianisation it was moved to Candlemas, 2 February (Monsen 1932: 280-281). This is probably a misunderstanding because it happened to be full moon on the evening of 1 February 1219, which means that the Disting market should have started on Candlemas that year. The fact that it was full moon on 1 February in 1219, which corresponds to 8 February in our calendar, and that the next year, 1220, was one of the years in the ancient eight-year cycle means that in 1220 the eight-year cycle was shifted from the end of February to its earliest date, 28 January

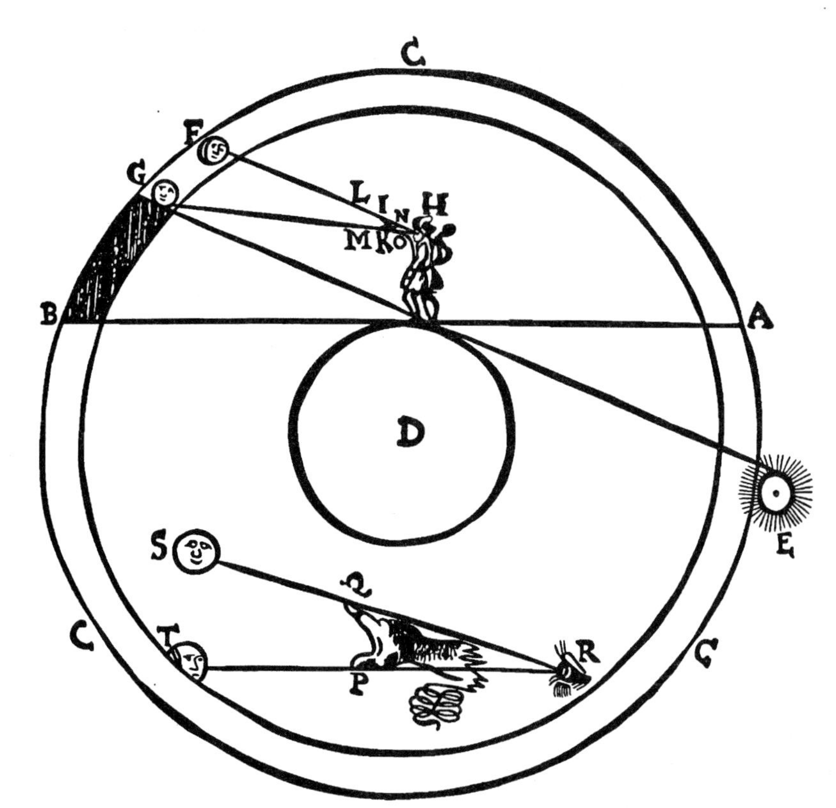

Fig. 2. Uppland farmers determined the dates of the next full moon by measuring the moon's distance at sunset from the "pale yellowish-red" night-ring. The distance between the thumb and the forefinger was called a 'span' and corresponded to the moon's movement during a 24-hour period (cf. Fig. 3).
E = the sun,
D = the globe,
the line A-B = the horizon,
the arc BG = the earth's shadow,
G = the night-ring at full moon,
F = the moon's position before full moon.
From Rudbeck 1689: 554.

The phases of the moon fall one day earlier after 304 years in the Julian calendar
According to Magnus (Foote 1996: 73-74 [Magnus 1.34]), the farmers of Uppland knew about the shifts between the true phases of the moon and the predictions in the Julian calendar during periods of 10 and even 600 or 1000 years. This knowledge must have been based on observations that covered at least one millennium before 1555, when his book was written. He described the runic calendar staffs, which had enabled the farmers from very early times to keep track of the calendar. There is a woodcut in his book (frontispiece in this volume) showing a mother instructing her daughter and a father his son in the art of using the calendar staff (see also Hallonquist in this volume).

The older people of Uppsala knew that in 1689 the Disting would fall for the first time one day earlier than it had during the last 300 years. When Rudbeck (1689: 652-653) asked the 90-year-old farmer why, he answered that there exists an old rule according to which the Disting's full moon had now completed the cycle of Aun and that it should be shifted by one day every 19th year during 300 and some years. He explained that the rule was called after the ancient King Aun and that 1689 was the correct year to adjust the date of the Disting, as one of his forefathers had engraved a half moon on his rune staff on the day of the Disting, the year, and the golden number of that year, and now 300 and some years had elapsed (Rudbeck 1689: 652). The statement of the old farmer as to the reason for the shift by one day in 1689 can be checked. I realized that the marking of the Disting full moon as a half moon in 1385 (1689-304 years) could be interpreted as the result of a lunar eclipse. This hypothesis could easily be verified because in 1385 the Disting's full moon was totally eclipsed after midnight and could be the seen as a half moon about 7 o'clock in the morning.

Shifts in the date for the Great Midwinter Sacrifice every eighth year according to the 304-year rule
During the 19-year cycle that began in 1689, the dates of the Disting full moon were shifted one day earlier. The first year to be shifted in the old eight-year cycle for the heathen Great Midwinter Sacrifice was 1692. This means that the earlier shifts had taken place in 1388, with no sacrifice, in 1084 with the last sacrifice, and in 780 and 476 with sacrifices. The year 476 falls within the estimated time of rule for the semi-legendary King Aun, 450–500. This is very interesting as King

Aun, according to the tradition preserved by the old farmer mentioned above, established this 304-year rule. This implies that the Julian calendar was introduced in Uppsala by at least 476 or that the ancient Swedes had independently invented a calendar with intercalation of one day every fourth year.

According to Lindqvist (1955: 79), the excavator of Old Uppsala, King Aun may be buried in the middle and oldest of the three burial mounds. In 476, it was full moon on the last day of the Disting period, when the sun was setting on the top of the middle mound. The excavations of these mounds showed that the burial ceremonies had great similarities with the cremation of Roman emperors, and Roman grave goods were found.

The Historian of Religion Ingemar Nordgren (2000: 1-2, 103-141) has used the eight-year cycle to reconstruct the cult of *Freyr* and the 304-year cycle of Aun to date the arrival of the cult connected with *Odin* in the Mälar valley region.

The latest shift in the 304-year cycle occurred in 1996 (1692+304 years), on 3 February, which corresponds to 21 January in the Julian calendar (Fig. 3).

Fig. 3. The rising full moon from the Tingshög (Thing mound) on 3 February 1996 at 16:13 Swedish Standard Time. The upper edge of the sun had disappeared below the horizon five minutes earlier, and the dark shadow of the earth's surface is clearly visible on a colour photo at half the altitude of the moon. The brighter part of the sky in this greyscale photograph is blue. The moon is visible in the "pale yellowish red" night-ring, between the blue sky and the shadow of the earth, which means that it was full moon in accordance with the rule used by the Uppland farmers in the 17th century (Rudbeck 1689: 649-661). Cf. Fig. 2. This method is much more exact for determining the phases of the moon than the observations of the new moon's crescent used by the ancient civilisations of the Mediterranean region and the Middle East. It was possible to determine without difficulty the date of a full moon and to mark the 19 different moon series on the calendar staffs. By comparing the markings on the staff with the observed dates of full moons, it should have been possible, without encountering any great problem, to discover the moon's displacements by one and a half days during the eight-year cycle and by one day after 304 years. The full moon on this photo is the latest in the series of 304-year cycles that include 1692, 1388, 1084, 780, 476, 172, and 133 BC and 437 BC.

Early Contacts between Sweden and the Greek and Roman world

In 476, the Western Roman Empire fell. A worn Roman coin from that year was found in the so called Ottar's mound in Vendel, ca 30 km north of Uppsala. The size of that mound is comparable to the mounds in Uppsala.

In 471, Theodrik the Great became king of the eastern Goths and resided at Ravenna. His name is mentioned on several Swedish runic stones, and on the most famous one, *Rökstenen*, in the province of Östergötland, the text tells us proudly that one of the ancestors had fought with Theodrik the Great.

If we continue some 304-year cycles before 476, we get AD 172, 133 BC, and 437 BC. If the ancient King Aun based his 304-year rule on Swedish observations, they must have started at least in AD 172. At that time the Gothic people living at the fall of the river Vistula into the Baltic immigrated to the Black Sea where there were old Greek colonies.

According to Ginzel (1911: 390), Hipparchos (d. 126 BC) introduced a 304-year period consisting of four 76 year cycles (introduced by Kallippos), or sixteen 19-year cycles (introduced by Meton), after which the phases of the moon will fall one day earlier. His reference for this in Ptolemy, however, is incorrect, and there is no other reference of this kind in the *Almagest*; nor do we know elsewhere of the use of such a period connected with Hipparchos. It is not mentioned by Geminos who lived 100 years after Hipparchos.

According to Diodorus (Oldfather 1961: 447-449 [Diodoros 12.2-3]), Meton is said to have introduced the 19-year cycle in Athens at the summer solstice in 432 BC. Hipparchos, who observed the solstices and equinoxes 146-126 BC, could in principle have observed and discovered this 304-year cycle in 133 BC, based on observations by Meton in 437 BC.

In any case it seems very unlikely that the early Swedes got the 304-year cycle from Hipparchos, not only because we know of no scholar in the Classical period who had any knowledge of this cycle, but mainly because the epoch for such cycles at that time was related to the equinoxes and solstices, whereas the epoch for the early Swedish 304-year cycle was the Disting period, determined by the second full moon after the winter solstice.

The heathen cult centre at Old Uppsala

In 1986, I noticed that the three burial mounds had been carefully aligned. The orientation corresponded to sunset on the date, 8 February, which regulated the restart of the eight-year cycle for the above-mentioned periodic sacrifice of humans and domestic animals at the Midwinter, or the Disting's, full moon, according to the lunar eight-year cycle. Midwinter month is the second month after the winter solstice and Midwinter Day is the day exactly between the winter solstice and the vernal equinox, according to Rudbeck (1679: 70-71), which means 2-3 February in prehistoric times.

The three mounds are also oriented in the direction of sunset on 3 November. In ancient Sweden, the year started at the first new moon after the 14 October, the first winter day on the runic calendar staff, or 21 October according to the modern calendar. This means that the first full moon of the year could earliest appear on 4 November in the modern calendar, only one day after sunset in the direction in which the northern sides of the three mounds were oriented. This difference of one day is insignificant, because the dates of the full moons were computed according to rules preserved on the calendar staffs, with the full moon marked only at 19 fixed dates within each month.

Three lunar months later, on 8 February, the sun sets again in the same direction. Another three lunar months later, on 29 April, the sun rises in the opposite direction. Every 19th year the moon will be full on all three of these days. The first day of the Disting period, 28 January, may have been defined by sunset at the top of *Tunåsen* (Tuna Ridge), the highest natural hill in the otherwise flat landscape, observed from an upright stone (menhir) on the Tingshög, the fourth large, but flat, mound. On the last day of the Disting period, 26 February, the sun sets on top of the originally smaller, middle mound, the oldest of the three 'royal' mounds. This mound has been dated by Lindqvist (1955: 79) to AD 450-500 and may be the tomb of the semi-legendary King Aun, who is believed to have reigned sometime during this period (Figs. 4-7).

The same dates were already important in the Neolithic calendar that was marked by grooves in the bedrock on the island of Gotland in the Baltic, by the orientation of the passage graves in Västergötland, and the interpretation of the calendar ships on the Swedish rock carvings (Henriksson 1983, 1989ab, 1992, 1994, 1995, 1999 and 2002). According to my interpretation the oldest grooves on Gotland were made on 27 January 3294 BC. The passage graves in Västergötland can be dated to ca 3300 BC, according to the archaeologist Lars Bägerfeldt (1989).

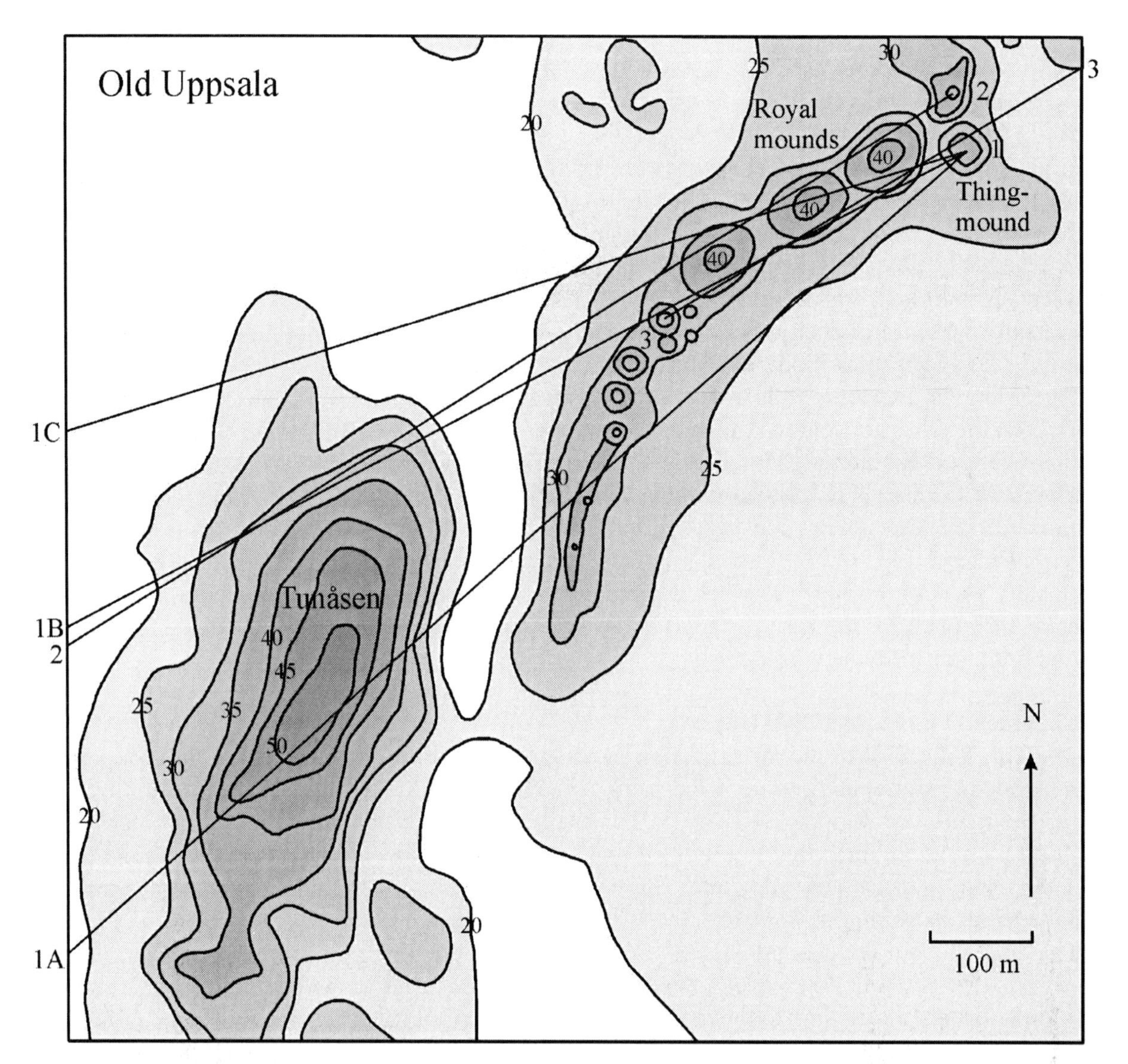

Fig. 4. Plan of the area occupied by the ancient monuments at Old Uppsala, with Tunåsen (Tuna Ridge) on the left. The three large mounds are situated largely on the line of the natural ridge, but their exact location, orientation, and design were determined according to a definite plan. From the menhir on the Tingshög (point 1, top right), one can see the sun set at the highest point of Tunåsen on 28 January, in a V-shaped depression between the northern slope of Tunåsen and the southern slope of the west mound on 12 February and between the crests of the west mound and the middle mound on 26 February (see Figs. 1-3). On the left-hand edge, the corresponding lines of sight are marked by 1A, 1B, and 1C.

From the smaller observation mound (point 2, top right), one can see the sun set in line with the northern sides of the east and west mounds on 3 November and 8 February, but the view is today partly blocked by trees (see Fig. 5). The line of sight has been marked with the number 2 on the left-hand edge. The rising of the sun in the opposite direction on 29 April and 13 August can, however, be observed without hindrance from the mound at point 3, to the left of the west mound (see Fig. 6). This line of sight has been marked with the number 3 on the right hand edge.

By observing the rising and setting of the sun along the sides of the mounds, a perfect control was established over the course of solar year and it was possible to determine the dates of the three important sacrifices at the beginning of the winter, the middle of the winter (midwinter) and the beginning of the summer.

Fig. 5. The sun set along the mounds on 3 November and 8 February. If there were a full moon 8 February, the annual midwinter sacrifice should be begun, but it was even more important to know that, if the great eight-year sacrifice were to be held the following year, this should be begun as early as 28 January. This photograph was taken on 8 February 1993 at 16:17 from the observation mound at the southeastern corner of the churchyard wall. The trees have been removed by the computer.

Fig. 6. If there were a full moon at sunset along the mounds at the beginning of the winter on 3 or 4 November, there would also be a full moon six lunar months later on 29 or 30 April, when the sun rose along the mounds, and then the sacrifice for the summer's battles was held. The photograph was taken on 29 April 1988 at 03:56 Swedish Standard time. By observing the sunrise and sunset along the mounds, it was possible to maintain a correct calendar.

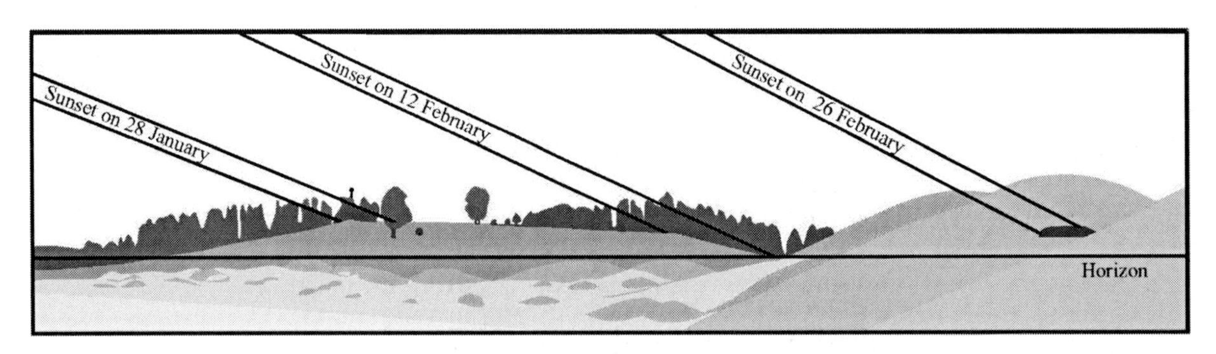

Fig. 7. The dating limits for the beginning of the midwinter sacrifice (later the Disting) could be established by observing sunset from the menhir on the Tingshög. On the first date, 28 January, the sun set straight over the highest point on Tunåsen and on the last date, 26 February, at the position of the burial cairn in the middle mound (Aun's mound?). These dates are given in our calendar. Every eighth year a Great Midwinter Sacrifice was made at the full moon, which fell between these dates. For each such major midwinter sacrifice the date of the full moon was moved forward by, on the average, one and a half days and, after 144 or 152 years, the displacement amounted to a whole lunar month, at which time counting began again from 28 January. In AD 468, there was a full moon on 26 February, when the sun went down where the burial cairn in the middle mound now stands, and in AD 476, it was time to re-start the cycle of sacrifices on 28 January at sunset over the highest point of Tunåsen. Even as late as the 17th century, the ancient King Aun was regarded as the inventor of the rule that predicted the displacement of the moon's phases by one day after 304 years. According to the interpretation of *Heimskringla. History of the Kings of Norway*, by Snorre Sturlason, Aun may have reigned in Uppsala about AD 450-500 and been buried there (Monsen 1932: 16-18, 22 note 5).

*Department of Astronomy and Space Physics, Uppsala University, Box 515, SE-751 20 Uppsala, Sweden, goran.henriksson@astro.uu.se.

Referenser

Aujac, G. (ed.) 1975. *Geminos. Introduction aux Phènoméns*. Paris.

Bägerfeldt/Blomqvist, L. 1989. *Megalitgravarna i Sverige* (Theses and Papers in Archaeology 1). Stockholm.

Foote, P. G., (tr.) 1996. *Olaus Magnus. A description of the northern peoples 1555*, vol. 1. London.

Ginzel, F. K. 1911. *Handbuch der Mathematischen und Technischen Chronologie,* vol. 2. Leipzig.

Granlund, J. 1958. *Kulturhistoriskt lexikon för nordisk medeltid*, vol 3, s. v. *Disting*. Malmö.

Henriksson, G., 1983. Astronomisk tolkning av slipskårorna på Gotland. *Fornvännen* **78**, 21-28.

Henriksson, G., 1989a. Archaeoastronomy. *Annual Report for 1988* (Uppsala Astronomical Observatory Report 49), 7. Uppsala.

Henriksson, G., 1989b. De västgötska gånggrifternas samband med solkult, Falbygden Årsbok **43**, 69-88.

Henriksson, G., 1992. Riksbloten i Uppsala. *Gimle* **20**, 14-23.

Henriksson, G., 1994. *Arkeoastronomi i Sverige*. Literature for course in archaeoastronomy at Uppsala University. Uppsala.

Henriksson, G., 1995. Riksbloten och Uppsala högar. *Tor* **27:1**, 337-394.

Henriksson, G., 1999. Prehistoric constellations on Swedish Rock-carvings. In *Actes de la Vème conférence de la SEAC, Gdańsk, 5-8 septembre 1997* (Światowit supplement series H: Anthropology, 2), ed. A. Le Beuf and M. Ziółkowski,155-173. Warsaw.

Henriksson, G., 2002. The grooves on the island of Gotland in the Baltic sea: a Neolithic lunar calendar. In *Proceedings of the conference "Astronomy of Ancient Civilizations" of the European Society for Astronomy in Culture (SEAC) and National Astronomical Meeting (JENAM), Moscow, May 23-27, 2000*, ed. T. M. Potyomkina and V. N. Obridko, 72-77. Moscow.

Hutton, M. (tr.) 1970. *Tacitus: Agricola. Germania. Dialogue on Oratory.* (Loeb Classical Library 35). Cambridge MA and London.

Lindqvist, S. 1936. *Uppsala högar och Ottarshögen*. Stockholm.

Lindqvist, S. 1955. *Uppsala och Jelling*. Copenhagen.

Monsen, E. (ed.) 1932. *Heimskringla or the lives of the Norse kings* (written by Snorre Sturlason ca 1230). Cambridge.

Nordgren, I. 2000. The possible origin of the Goths traced through the symbolism of rings and through place-names containing the element *ring*. *Migracijske teme* **16**, 1-2: 103-141.

Nyberg, T. 2000. *Monasticism in North-Western Europe, 800-1200*. Aldershot.

Oldfather, C. H. (tr.) 1961. *Diodorus of Sicily*, vol. 4 (Loeb Classical Library 375). Cambridge MA and London.

Reuter, O. S. 1934. *Germanische Himmelskunde*. Munich.

Rieper, H. (ed.) 1995. Ansgar und Rimbert: die beiden ersten Erzbischöfe von Hamburg/Bremen und Nordalbingen. Hamburg.

Rudbeck, O. 1679a. *Atlantica*, vol. 1. Uppsala. 1937 reprint.

Rudbeck. O. 1679b. *Atlantica*, plates. Uppsala. 1938 reprint.

Rudbeck, O. 1689. *Atlantica*, vol. 2. Uppsala. 1939 reprint.

Sundqvist, O. 2002. *Freyr's offspring, Rulers and religion in ancient Svea society* (Acta Universitatis Upsaliensis. Historia Religionum 21). Uppsala.

Toomer, G. J. (tr.) 1984. *Ptolemy's Almagest*. London.

Tschan, F. J. (tr.) 1959. *History of the Archbishops of Hamburg-Bremen* (written by Adam from Bremen ca 1075). New York.

Trillmich, W. (tr.) 1957. *Chronik av Thietmar von Merseburg.* Darmstadt.

The Twelve Days at Stonehenge

Stanislaw Iwaniszewski*

Abstract
The particular alignment of the Heel Stone at Stonehenge is discussed in light of the Twelve Days hypothesis. It is argued that the location of this stone served to mark intervals of 8-11 days, possibly related to lunisolar calendars and festivities pivoting about solstice dates.

Introduction
Over the last forty years archaeoastronomers have revisited, reassessed, or dismissed many of the supposed astronomical alignments at Stonehenge. While it became clear that the so-called Aubrey holes could in no way serve to predict eclipses and that many of the lunar alignments and other claims made for the monument were proved to be non-existent, the astronomical significance of the Heel Stone has not been explained in a satisfactory way. Its location as viewed from the centre of the Sarsen Circle has been considered to be a crude midsummer sunrise alignment.

In the 1960's and 1970's, Stonehenge was regarded as a precise astronomical observatory with alignments coinciding with solstitial solar and lunar orientations. There have been also claims for the use of the monument to predict eclipses (Hawkins 1963, 1964; Hawkins and White 1965; Hoyle 1966a, 1966b; Newham 1966), while the discrepancy in the orientation of the Heel Stone needed an adequate explanation. If Stonehenge is supposed to have been planned as an astronomical observatory, then it was built as accurately as possible, and the Heel Stone's deviation from the solstitial alignment was done deliberately.

In this vein Hawkins (1964: 1258) was the first to observe that the full moon at the winter solstice could have risen over the Heel Stone just before eclipses. This idea was accepted by Atkinson (1966: 215), and scholars started to believe that this was the best explanation for the particular placement of the Heel Stone. A more radical position was adopted by Robinson (1970) who regarded the stone as an exclusive winter eclipse moonrise indicator. Robinson argued that although the declination of the moon at this time was the same as that of the sun at the summer solstice, it rose to the right of the sun due to the parallax effect. The idea of a deliberate and precise alignment of the Heel Stone was reassessed.

During the 1970's, the concepts of the function of the Heel Stone remained unchanged (e.g. Lancaster Brown 1976: 46-48, 62-63; Krupp 1979: 116-119; MacKie 1977a: 112; 1977b: 120-122; Wood 1978: 163). In 1979, rescue excavations directed by Pitts (1981; 1982) unearthed the area around the Heel Stone and revealed a pit, presumably a stone hole where a now-missing companion to the Heel Stone had been deposited. This discovery was followed by a number of investigations that supported a midsummer sunrise hypothesis (e.g. Atkinson 1982; Hawkins 1985; Ray 1987). According to this hypothesis both stones, the missing stone 97 and the Heel Stone, formed a frame in the centre of which midsummer sunrise could have been observed. The lunar theory has been abandoned.

Already in the 1970's, some scholars noticed that the sun was seen over the Heel Stone some days before and after midsummer sunrise. The exact number of days varies according to the different chronologies for the placement of the Heel Stone: MacKie (1977b: 121) gives a number of eight days, assuming that the stone was set up around 2800 BC; Ray (1987: 252) counts nine or ten days for the date of 3100 BC and argues that the position of the Heel Stone was chosen to mark the "beginning and ending of a ritual period around the time of the summer solstice".

The Twelve Days at Stonehenge
The azimuth of the peak of the Heel Stone as seen from the centre of the Sarsen Circle is found to be 50°54', and its altitude is 47'29" (Atkinson 1978: 51-52). Since, as Atkinson writes (1978: 52), the exact value of (the original) horizon elevation cannot be established more precisely, and refraction can vary from day to day according to local climatic conditions, the results of any calculations must be seen as approximate. As the mean refraction is about 27'45"—corrected for the altitude of Stonehenge above sea

level, it is around 27'—it results that the sun's declination can vary between 23°23' for the first gleam of the sun over the top of the stone and 23°34' for the sun's centre to emerge above the stone (Table 1).

Alignment	Azimuth	Horizon elevation	Declination	First gleam	Sun's centre
Centre towards Heel Stone	50°54'	0°47'29"	23°57'	23°45'	23°31'

Table 1. Basic data at Stonehenge.

According to the data displayed in Table 2, the sun's declination at the summer solstice diminished 4 arc minutes during the period of 700 years. This gives the difference in the sun's declination of 17 to 13 arc minutes for the first gleam, or of 31 to 27 arc minutes for the centre of the sun's disk.

Epoch	Sun's Declination
-3100	24°02'06.98"
-2800	24°00'26.71"
-2500	23°58'41.72"
-2400	23°58'05.73"

Table 2. Declination of the sun at particular epochs.

Assuming that in 2001 AD, the sun changes its declination near the solstices with more or less the same rate per day as some 5000 years ago, then the difference of 13'-17' corresponds to the period of 8-9 days before and/or after the solstice, while the difference of 27'-31' equals the period of 11-12 (13) days before and after the solstice. In 2001 AD, the summer solstice would fall on June 21, its declination would reach the margin of 13'-17' on June 12-13 and July 29-30, while on June 9-10 and July 3-4, it would be 27'-31' off its midsummer position (Table 3).

Date in 2001 AD	Declination	Solar events
June 9	22°55'02.5"	31' limit
June 10	22°59'53.5"	27' limit
June 12	23°08'22.7"	17' limit
June 13	23°12'00.7"	13' limit
June 21	23°26'17.3"	summer solstice
June 29	23°14'09.5"	13' limit
June 30	23°10'47.6"	17' limit
July 3	23°58'16.4"	27' limit
July 4	22°53'17.8"	31' limit

Table 3. The 8 and 9 and 11 and 12-day periods around the summer solstice according to *The Astronomical Almanac for the Year 2001.*

The position of the Heel Stone marked the intervals of 8-9 days before and after the midsummer solstice with reference to the first gleam of the sun and the intervals of 11-12 days before and after the midsummer solstice with reference to its centre. The observer at the centre of the Sarsen Circle could have observed the rising of the sun over the Heel Stone at intervals of 8-9 or 11-12 days before and after the midsummer solstice for most of the 3rd millennium BC. As the Heel Stone was probably installed around 2400 BC, it gives the intervals of 8 and 11 days respectively.

The possible meaning of the Twelve Days

If we admit that the missing stone 97 and the Heel Stone were deliberately placed to form a frame in the centre of which a midsummer sunrise could have been observed, then we can also admit that the position of the Heel Stone could have been chosen to mark those 8-9 or 11-12-day periods. So instead of admitting the existence of a crude midsummer sunrise alignment, we can ask about the possible significance of the periods of 8 or 11 days.

A good summary of the theories concerning the Twelve Days (or Twelve Nights) is found in Van Gennep's monograph on French folk customs related to this interval of time. Van Gennep (1958: 2860-2863) identifies three theories explaining the origin of the period of Twelve Days. These are based on: (a) both solstices, (b) his theory of the *rites de passage*, (c) mortuary practices and funeral festivities. Frazer (1919: 325-326) was one of the first scholars to argue that the interval of the Twelve Days, widely known in folk customs all over Europe, probably reflected the old intercalary period during which the difference between 12 lunar months and a solar year had been observed. Nilsson (1920: 306) cites a Finnish example establishing the rule that when the "first heart-moon is born late e.g., after the Twelfth Day, there is no second heart-moon in this year".

In a lunisolar calendar, one revolution of the sun (adopting a geocentric position) must be kept in accordance with the revolutions of the moon. Actually, the length of the solar (tropical) year is 365.242199 days. This length must comprise 12 (12 x 29.530589 = 354.3671 days) or 13 (13 x 29.530589 = 383.8977 days) lunar (synodic) months. In order to keep lunar time reckoning in accord with the length of the solar year, it is necessary to introduce an occasional month into the series of 12 lunar months. So a year of 13 lunar months must appear from time to time instead of that of 12 months, to keep both calendar systems together. One lunar year of 12 months is about 11 (365.242199 - 354.3671 = 10.8551) days shorter than a solar (tropical) year. After two years of 12 lunar months, this difference accumulates to 21.71026 days, and after three such years, it will be greater (32.565442 days) than the length of a (synodic) month. It is a proper time to proceed with the intercalation of a month (McCluskey 1977: 183).

In 1700 BC, the length of a tropical year was 365.24242 days and that of a synodic month 29.53060 days (Thom 1971: 117); so the difference between the two was 10.87522 days (12 lunations = 354.3672 days), which is essentially the same as it is today.

Since the midsummer date may be regarded as a kind of turning point, it could serve as a good indicator for deciding whether or not to insert an additional month. The people of Stonehenge could have been interested in observing successive sunrises over the Heel Stone 11 days before and after the summer solstice, regarding this period as a 'window' for observing lunar phases.

Let us proceed with an example. Assume that in the first year the rising of the sun over the Heel Stone, as it turns southward just after the solstice, coincides with the new crescent moon. A year later the thirteenth new moon will coincide with the date of midsummer sunrise. In another year, the 25th lunar month starts when the sun appears over the Heel Stone in its northward movement, just before the solstice. In the fourth year, the 37th lunar month will start another 11 days earlier, while the sun is still far away from the Heel Stone. This may indicate that an intercalary month should be added. In other words, while the sun is seen on the left side of the Heel Stone and the new crescent appears in the western sky, there is no need to proceed with intercalation; when the new moon appears and the sun is still on the right side of the Heel Stone, there is the need to add a month.

It has been argued that lunar alignments dominated at Stonehenge I (Ray 1987: 251, 253-4; Bradley 1991: 215; Burl 1987: 65). If so, it is probable that a lunar time-reckoning system was well developed. The Stonehenge solstitial alignments may reflect the development of a new time-reckoning device based on the sun. The existence of two types of time-reckoning principles might have resulted in the development of a lunisolar calendar.

A possibly similar function is displayed at other megalithic monuments, e.g. at Newgrange and Woodhenge. At Newgrange, in a huge passage grave dated to about 3100 BC, the rays of midwinter sunrise enter a passage to illuminate the central burial chamber. The roof box over the entrance to the passage is skewed, allowing the sun's rays to enter "about a week before and a week after" the winter solstice (Ray 1987: 253). Patrick (1974: 518) calculated that the roof box is opened towards declinations between -25°53' and -22°53'. Allowing that the sun's declination at the winter solstice in 3100 BC was 24°02'06.98", there is a difference of about 69'. So in 3100 BC, the sun's rays could have entered and left the passage some 17 to 18 days before and after this event. Therefore the sun would have illuminated the passage for quite a long period of some 35 days; observe that this calculation is not precise. On the other hand, after three years the difference between a lunar year and a solar year accumulates to 32.5 days. So when the proper lunar crescent did not appear in the sky while the sun's rays were still inside the chamber, a new intercalary month should have been added. In this system lunisolar intercalation was pivoted about the winter solstice.

Examples of different lunisolar calendars suggest that when intercalation was done regularly, the additional month was placed at the end of the year, just before the beginning of the new year (Nilsson 1920: 277). One of the best examples of coordinated lunar and solar time reckoning is preserved in the so-called Calendar of Coligny. For the purposes of the present paper it is not important whether the calendar is of Gaelic origin or a Roman reconstruction of a Gaelic calendar. This calendar covers a span of 62 months of 29 or 30 days, representing 5 years divided into two halves, starting on the 1st of November and the 1st of May respectively. In order to keep the periods of 12 lunar months in step with the solar year, after each period of 30 lunar months (2.5 years), an additional lunar month was added. These intercalary months were added at the solstices. In other words, when the intercalary month of 30 days was added before Samon (summer solstice), 2.5 years later a new intercalary lunar month was added before Giamon (winter solstice) (Ljungman 1938: 444-446). For Ljungman (1938: 447) this could have been also the case in Irish, Welsh, some German and Brittany calendars, and the additional intercalary month started "mit dem letzten Neumond *unmittelbar vor einem der Solstitien*".

In the British Isles, the old Anglo-Saxon lunisolar calendar described by the Venerable Bede allowed for the intercalation of a thirteenth month in the summer season (Nilsson 1920: 295; Harrison 1973: 285). Even if Bede's statements on the beginning of an Anglo-Saxon year are unclear, it is evident that the intercalation of a third Litha month (meaning 'moon' according to Harrison) took place in the summer, possibly after the solstice.

These examples cannot, of course, trace calendar systems back to Neolithic times, but they show us an ancient tradition of intercalating an additional lunar month around the solstices. On the other hand, McCluskey (1989) concluded that the division of the year into halves and quarters could have been found in orientations of the British megaliths. In light of the evidence discussed here, I suggest that the traditional intercalary system pivoting about the solstices may, in fact, have originated in the Neolithic.

Localised in the landscape previously ordered by Neolithic long barrows emphasising particular locations "as special, sacred, or associated with the activities of ancestors and divinities" (Thomas 1996: 196), the series of Stonehenge enclosures marked the change of societal orientation. Earlier stone and earthen long barrows accentuated burial and funerary practices, providing symbols formerly given by domestic architecture (long houses) and, especially, the dimension of durable visibility. The use of enclosed spaces for ritual activities marked a new pattern of defining areas of space; the series of ditches and enclosures apparently put stress upon the boundaries—the bounded area being separated from the outside world (Thomas 1996; Bradley 1998: 126-130). As they were predominantly mortuary monuments, the separation of the dead from the world of the living has been suggested. Around 2500-2400 BC, the rituals performed within the Stonehenge enclosure lost their mortuary connotations. Earlier timber structures in the centre were replaced by more durable structures made of stone, and the Heel Stone was installed (e.g. Bradley 1998: 93-94). The visual sight lines between the monument and its surroundings were extended to the far horizon, uniting different places in a kind of sacred or ritual geography. While earlier stone monuments could have emphasised specific links between the dead and the sky, after 2400 BC it was the lunisolar calendar bringing space and time together. In later times the intervals of 8-11 days became ritualised, denoting the length of festivities carried out on this occasion (Vaiškunas, this volume) or the period of 'sun-dancing'.

Conclusions

In light of the investigations of McCluskey (1989), Frazer (1919: 325-326), and Ljungman (1938: 428-431, 434-438), it is reasonable to suggest that the Stonehenge calendar started around the solstices and that the Heel Stone was deliberately placed to mark the 11-day intervals around the summer solstice. This is a much better explanation for the Heel Stone.

In archaeoastronomical analysis human societies seem irrelevant either to the development or functioning of Stonehenge. Archaeoastronomers have viewed the monument as a physical phenomenon built to bear the presumed solstitial alignments. When less precise solstitial alignments were found and the function of the Heel Stone as an solstitial marker was dismissed, they started to look for another pit which might have held a stone indicating the approximate point of solstitial sunrise. Selectively excluding other astronomical-calendrical possibilities, they missed the possibility of interpreting the location of the Heel Stone in terms of the Twelve Days interval. In light of the present hypothesis the location of the Heel Stone is not wrong, but purposeful and meaningful.

*Division de Posgrado, Escuela Nacional de Antropologia e Historia, c. Periferico Sur y Zapote s/n, Deleg. Tlalpan, 14030 Mexico, D. F., siwanisz@yahoo.com.

References

Atkinson, R. J. C. 1966. Moonshine on Stonehenge. *Antiquity* **40**, 212-216.
Atkinson, R. J. C. 1978. Some new measurements on Stonehenge. *Nature* **275**, 50-52.
Atkinson, R. J. C. 1982. Aspects of the archaeoastronomy of Stonehenge. In *Archaeoastronomy in the Old World*, ed. D. C. Heggie, 107-116. Cambridge.
Bradley, R. 1991. Ritual, time and history. *World Archaeology* **23**, 209-219.
Bradley, R. 1998. *The significance of monuments. On the shaping of human experience in Neolithic and Bronze Age Europe.* London and New York.
Burl, A. 1987. *The Stonehenge people.* London.
Frazer, J. G. 1919. *The golden bough. A study in magic and religion.* Part 4: *The scapegoat*, 3rd ed. London.
Harrison, K. 1973. The primitive Anglo-Saxon calendar. *Antiquity* **47**, 284-287.
Hawkins, G. S. 1963. Stonehenge decoded. *Nature* **200**, 306-308.
Hawkins, G. S. 1964. Stonehenge: a Neolithic computer. *Nature* **202**, 1258-1261.
Hawkins, G. S. 1985. Stonehenge archives: Twenty years after "Stonehenge decoded". *Archaeoastronomy* (Bulletin of the Center for Archaeoastronomy) **8**, 6-9.
Hawkins, G. S. and White, J. B. 1965. *Stonehenge decoded.* New York..
Hoyle, F. 1966a. Stonehenge—an eclipse predictor. *Nature* **211**, 454-456.
Hoyle, F. 1966b. Speculations on Stonehenge. *Antiquity* **40**, 262-276.
Krupp, E. C. 1979. *In search of ancient astronomies.* London.
Lancaster Brown, P. 1976. *Megaliths, myths and men: An introduction to astro-archaeology.* Poole.
Ljungman, W. 1938. *Traditionswanderungen Euphrat-Rhein. Studien zur Geschichte der Volksbräuche*, vol. 2 (*Folklore fellows communications* 119). Helsinki.
MacKie, E. 1977a. *The megalith builders.* Oxford.
MacKie, E. 1977b. *Science and society in prehistoric Britain.* London.
McCluskey, S. C. 1977. The astronomy of the Hopi Indians. *Journal for the History of Astronomy* **8**, 174-195.
McCluskey, S. C. 1989. The mid-quarter days and the historical survival of British folk astronomy. *Archaeoastronomy* (Supplement to the *Journal for the History of Astronomy*) **13**, S1-S19.
Newham, C. A. 1966. Stonehenge: a Neolithic "observatory". *Nature* **211**, 456-458.
Nilsson, M. P. 1920. *Primitive time-reckoning* (skrifter utgivna av Humanistiska Vetenskapssamfundet 1). Lund.
Patrick, J. 1974. Midwinter sunrise at Newgrange. *Nature* **249**, 517-519.
Pitts, M. W. 1981. The discovery of a new stone at Stonehenge. *Archaeoastronomy,* (Bulletin of the Center for Archaeoastronomy) **4**, 17-21.
Pitts, M. W. 1982. Stones, pits and Stonehenge. *Nature* **290**, 46-47.
Ray, B. C. 1987. Stonehenge: A new theory. *History of Religions* **26**, 225-278.
Robinson, J. H. 1970. Sunrise and moonrise at Stonehenge. *Nature* **225**, 1236-1237.
Thom, A. 1971. *Megalithic lunar observatories.* Oxford.
Thomas, J. 1996. *Time, culture & identity: An interpretative archaeology.* London.
Van Gennep, A. 1958. *Manuel de folklore français contemporain*, vol. 1, 7: *Cycle des douze jours.* Paris.
Wood, J. E. 1978. *Sun, moon and standing stones.* Oxford.

Some aspects of Lithuanian folk observations of the sun during the summer solstice period

Jonas Vaiškūnas*

Abstract
Lithuanian folk traditions, as many others throughout the world, show that much attention was paid to the observation of the positions of the sun at the equinoxes and solstices. I shall concentrate here only on the beliefs and practices related to the summer solstice periods and, more especially, on the references to the 'dancing sun', the 'standing sun', and the 'returning sun' during the festival of Saint John and the calendar period related to it. These folk traditions are most certainly remnants of old-time practices of observations of the sunrise and sunset azimuths. I shall also present some comments on material collected from Lithuanian folklore from the 19th and 20th centuries.

Introduction
The festival *Rasa*, 'dews' or *Kupolės*, is a midsummer feast celebrated of old in Lithuania. At the time of the Christianisation of Lithuania, in the late 14th and 15th centuries, this pagan festival, which coincided more or less with the Christian feast of St. John the Baptist, survived as *Joninės*, 'John's feast'. This festival marks the celebration of the summer solstice. However the many traditions that have reached our time seem to pay little attention to the sun itself and concentrate rather on rituals related to fire, water, and flora, a fact often remarked upon by ethnographers and folklorists. For instance, in his survey of the Lithuanian customs of the *Joninės* festival, Balys notes that it is difficult to trace a clearer link with the cult of the sun and it's returning. The idea of the floral spirit is more present (Balys 1993: 240-242).

The analysis of ethnoastronomical material collected about the *Joninės* festival, however, shows that the solar aspects of this calendar celebration must have been much more important in olden times. The calendar system based on the annual cycle of the sun's motion must necessarily have led to the recognition of the two most remarkable positions of the sun and the associated dates. These crucial positions, which are also easiest to mark on the horizon, occur at the summer and winter solstices. Defining the dates and the ritual marking of the dates must have been of great importance in the traditional Baltic and Old Lithuanian cultures, in which the organisation of daily life was carried out in compliance with the solar and lunar calendars. Recently, growing interest in archaeoastronomical and ethnoastronomical research indicates that in archaic cultures the sun, the moon, and other heavenly bodies were carefully observed. In Lithuania the hill known under its popular name of the Birutė' Mountain, in the town of Palanga near the Baltic Sea, has been considered as an ancient ritual area for astronomical observations. According to the archaeological investigations, in the 14th and 15th centuries there was a set of wooden poles serving these observations (Fig. 1).

The poles would have permitted the observations of the rising and setting azimuths of the sun and the moon (Lovčikas 1996: 128-129, figs. 7-9). Memory of such practices has survived in Lithuanian folklore until our day; for example the text of the following song proves that the sun was observed with the help of wooden poles:

> On the sea on the wide blue
> The Sun was 'poling'.
> On two – three poles,
> On nine arrows.

The visible northern and southern limits of the motion of the sun on the horizon and the way they were observed have left their marks in rituals and oral traditions, which they have strongly influenced. I shall present here some of these remnants of the *Rasa* festivities from the archaeoastronomical point of view.

Fig.1. Reconstruction of a pole observatory on the Birutė Hill in the Palanga Sea. With permission.

The length and calendar location of the summer solstice festival

As one of the main goals of the astronomical observations was the determination of important calendar dates, let us look at the question of the date of the *Rasa* festival. According to Stryjkowski (1846: 146) these festivities started on 25 May and lasted until 25 June. Daukantas (1976: 541) holds that the *Rasa* festivities lasted 14 days. Until recently in neighbouring Latvia the rituals lasted for a whole week. But in other parts of Europe the dates were sometimes different. In Austria, for example, they are reported to have lasted for about 12 days (24 June-4 July). Nowadays in Lithuania and in other European countries among peasants and country folk these festivities usually take place between 24 and 29 June. We may suppose that the broad range of dates covering this festival could be related to the observations of the changes in the azimuths of the rising and setting sun, its slowing down, and its standstill at the solstices preceding its return in the other direction. The fact that country people were well aware of the directions of the rising and setting sun, not in the east and west, but rather far in the northeast and northwest on the summer solstice days, and that they observed them carefully is shown by the expressions used for these times of the year. The locations of midsummer sunrise and sunset are referred to as the "high east and the high west of the summer" and the directions of winter sunrise and sunset as the "low east and low west of the winter". Some people still use specific places of the landscape to mark the location of the sun at its extreme. For example: "On the Seliutai (a landowner's family name) oak, the sun rises and it sets on the Pamociškes slope in February; and when the days are getting longer, then March comes. When the day gets longer, we say that the sun rises on the Kalnas hill. Every hill or elevation of ours has a name" (5).[1] Here the familiar features of the local surroundings, such as a tree on the neighbour's property, are used to mark a milestone in the calendar. If we accept a precision of about 1° for the measurements of the azimuth, the inhabitants must have considered that the midsummer sun had reached its extreme position and standstill between about 13 and 30 June, that is, a period of up to 18 days. The central date of this period corresponds to the astronomical solstice date. The word *solstice* is derived from the Latin *solstitium* meaning 'stand-still of the sun' and the same meaning is found in other languages, cf. the Russian term

[1] The numbers in parentheses refer to the names of persons in the list at the end of this paper.

letneje solncestojanije, 'the summer standing of the sun' or summer solstice. But in Lithuanian, the summer solstice is called *saulėgrįža*, 'turning round of the sun', where the process of the returning of the sun is emphasised rather than its standing still. We can easily accept that this 'turning round' means the changing of direction of the sun on the horizon around the northern limit of its azimuth, but it also indicates a turn in the length of days and nights. Until then the days were growing longer and the nights were becoming shorter, and after that time the proportions of the days and nights were reversed. We also find clear expressions of the observation of this standing and regression of the sun, e.g.: "The sun is standing for about two weeks on St. John's and afterwards the days begin to grown shorter until Christmas" (4). "When the day is long—the sun is standing at the same place for four days—and then the sun *springs* to its feet and the days grow shorter" (5). "The sun *stands in one place* from St. John's to St. Peter's and then the days get shorter". And they say about it that "the length of the day *jumps back*" (6) or, more often, "the sun *jumps back*".

It is possible that in the old times the summer solstice festival time was associated with the above-mentioned 'sun standing' period.[2] And the culmination of the feast might have been related to the end of the 'sun standing' period, when it is said that it 'jumps back', the so called 'turning back' of the sun. This supposition would allow us to explain why in historical sources the *Rasa* festival occupies such a long period around the solstice and also to understand the meaning of the expression 'dancing sun'.

Some possible explanations for the 'dancing sun'

It could be that such expressions as 'turning round of the sun', 'the sun jumps back', and 'the sun springs to its feet' first designated simply the movement of the sun on the horizon during the summer solstice period. But these words seem to describe the movements of a dancer: to spring to the feet, to jump back and forth, to turn around; and so, by contamination, people would have started simply to use the word 'dance' as well and to understand it with its strongest meaning. And then, as the phenomenon is common in folklore, these words would have soon been taken by simple minds to mean really that the sun was dancing. This belief that we can observe the sun dancing on the morning of midsummer night is common all over Europe; and in Lithuania, for instance, the dancing sun is called *Saulė šoka*: "The sun dances (*šoka*) on 29 June and jumps back (*atšoka*), and the days get shorte" (3). "On St. Peter's, the sun dances (*šoka*) and the days get shorter" (2). "On St. Peter's, the sun jumps back (*atšoka*). The sun jumps (*šokinėja*), the witches jump (*šokinėja*), and the sun. The length of the day jumps back (*atšoka*). From St. John's to St. Peter's it stops at the same place and after that the days get shorter" (6). "On St. Peter's day the sun dances (*šoka*) and shimmers. Perhaps then the sun jumps back (*atšoka*) and the days get shorter" (4). "On St. Peter's the sun jumps back (*atšoka*) and the days start visibly getting shorter. [...]. The sun returns back (*grįžta atgal*) and the days get shorter from St. Peter's..." (1). This association and confusion between turning and returning, jumping, springing, and dancing are thus frequent and well-documented. Yet most often it is simply said that the sun dances (*šoka*) at its rising. Sometimes it is also said that it dresses up or changes colours (*rėdosi*) (Balys 1993: 240-242). In popular thought 'the dance' of the sun is associated with the 'the jump', i.e., when the sun moves back and forth from its extreme rising and setting azimuth.

The motive of the dancing sun is broadly attested in Europe. Latvians, Byelorussians, Russians, French, Italians, Poles, Bulgarians,[3] and Greeks also mention the dancing ('playing', 'rolling over', etc.) sun on the mornings of St. John and St. Peter. In most areas it is said that the sun is dancing on the morning of St. John and St. Peter, but other sources mention, curiously, the same dancing of the sun for Shrove Tuesday morning or Easter morning. Could this mean that there is another reason for this belief than the one proposed first—a contamination of the turning, returning, standing, and jumping sun at the

[2] For example the Zuñi in North America celebrated midwinter when the rays of the rising sun struck a certain point on a certain mountain. They described the solstice as the time when the sun set for four successive days at the same point (Thurston 1994: 10).

[3] See information from the central Rhodopes in the article by Koleva in this volume.

solstices? It could be also that this dancing of the sun reflects an optical illusion due to the changing colours and azimuths of the sun on or near the horizon. The diffraction of light and the filter of atmospheric layers give a different glow to the morning sun, from dark red to pale white. This would explain the dressing up of the sun depending on different temperatures of the atmospheric layers. There is still another possibility: atmospheric refraction can cause us to see the sun above the horizon when, in fact, it is still under it. The phenomenon results from the fact that the rays of light follow the atmospheric layers of different temperatures and humidity, functioning as mirrors and carrying the light exactly in the same way as an optic fibre. Refraction always lifts up objects vertically. In northern latitudes, where the angle of the rising sun to the horizon is important, this can displace its azimuth in a significant manner and, if the changes in atmospheric conditions such as temperature and humidity of the air are important at the time of the solstice, the sun, instead of seeming to stand still for a few days, can give the illusion of stepping back and forth a few times before seeming more decidedly to move back. These effects of refraction could well explain why the phenomenon is so widely related to the feast of St. John and not other festive occasions such as Shrove Tuesday or Easter Sunday. So we can risk one more hypothesis: It is possible that at all larger festivities the habit was both to dance and to observe sunsets or sunrises, and again the proximity of these two activities would have directly led to an amalgam, and the 'dancing sun' would simply be the result of a linguistic contamination between these two activities.

There may be another possibility. The process of observing the rising and setting of the sun, whether or not for ritual reasons, implies staring at the sun for a long time. This produces a quasi-hypnotic state and subjective optical aberrations: the sun seems to tremble, jump from its place and turn around itself more and more quickly (the rotating suns of Vincent Van Gogh illustrate well this optical illusion).

All these accounts of the dancing sun at certain dates would simply mean that these were the only times of the year when the sun was stared at. They reflect only the exceptionality of these festivals and give us indications about the times in the year for the most prominent ancient pagan festivals.

At any rate, all these expressions show clearly that solar observations were practised of old in European pre-Christian traditions at certain crucial times of the year. They were carried further by the Christian cult at the same dates and very often at the same places. It is quite possible that the ritual pole of the *Rasa* festival, the *kupolė*,[4] could be used for such observational purposes. The *kupolė* could have been used as a sort of gnomon, casting a shadow to mark a ritual (Figs. 2, 3).

Fig. 2. Ritual poles of the *Rasa* festival. With permission.

[4] *Kupolė*, or *kopolis*, is a pole covered with bark, decorated with flowers and greenery (like a Maypole), and placed in the ground at the site of the feast, usually on a hilltop.

Fig. 3. Ritual pole of the *Rasa* festival. With permission.

We know from modern ethnographical materials that the use of a gnomon was well understood in folk traditions. Sometimes a full description is given, e.g. in the following fragment: "When the sun moved furthest north, the day would be longest. Then it is St. John's. " The longest day would be determined in the following way. "Take some tree in the middle of the fields or something alone that is not hampered by shadows. One day the length of the shadow of the tree in the sunset would be marked, the end of the shadow marked by a pole. On the next evening the procedure would be repeated. Not only the length of the shadow would be marked, but also the angle by which it falls from the tree. When the shadow of the tree reached the furthest point [to the south], it was considered to be the longest day" (5).

We can find in older Lithuanian ethnographical material the clear mention of the religious ritual of tree gnomons raised up only on certain specific occasions. On Shrove Tuesday there was the tradition of fixing a wheel on a high pole symbolising the sun of the St. John celebration *Joninės*. According to the length of the shadow that was cast by the pole, people counted time and determined the date for the festival of the longest day (Kudirka 1992: 30).

Other indirect attestations of the fact that the observations of the sun were carried out are the following: "Only those who sleep not on St. John's night can see the dancing and dressing sun". And it is said that this observation should be done through a silk scarf (Balys 1993: 241), through a smoked glass, or through a slot in the bathhouse (Balys 1935-1940). The last condition of observing sunrise from the bathhouse is most interesting. It recalls the technique of the camera obscura to mark the azimuth or declination of the sun known from other archaeoastronomical and ethnoastronomical sources. A calendar date is marked according to the short time that a sunbeam illuminates a special mark through a small hole in a dark place, a temple, or a cave. But we may have here a further indication of the sanctity of this period of the year in the pre-Christian Lithuanian culture, as the steam bath was at that time a ritual place used for purification before all great events of private and public life at all important times of the year. It must then be a pagan remnant. We know also that the Christian church fought against the custom of steam baths in Lithuania until as late as the 16th century.

Ethnographer J. Petrulis has recorded very interesting information concerning solstitial observation in Lithuania (Trinkūnas 2000: 72). A certain village craftsman, Butvila by name, would cease work and disappear for two weeks around the time of the summer solstice. Upon his return he would say, "A difficult year it would be for me if I were not to observe the return of the sun, as in the summer she presents herself. " The warmest time of the year is certainly favourable to isolation and, if the Christian patron of this time is Saint John the hermit, the folk traditions could have accepted it the more easily since everywhere it was the time of spending time in complete loneliness or with only sheep and cattle.

It is thought that observation of the heavenly bodies was one of the most important occupations of the wizards and priests in pre-Christian times. When Christianity became firmly established, the observation of the heavenly bodies and certain other ancient rituals were kept and performed only by senior members of the family. But when the observation practices were destroyed by the exclusive use of the computus for the establishment of feast dates, the astronomical traditions based on direct observation were shattered first. The flora and fauna motives survived for a longer time because they were less obviously bound to pagan religious beliefs; they were also more practical and more comprehensible to the people.

Conclusions

1) Lithuanian folklore and ethnographical material have retained remnants of the practices of observation of the sun in order to define calendar dates.
2) The length of the Lithuanian summer solstice festival was related to the observation of the so-called 'stopping' of the sun and the slow decrease and increase of the progression of its azimuths in the morning and evening.
3) The expression 'dancing Sun' must have its origin in these observational practices of the sun at the horizon.

Acknowledgements

The author wishes to thank the Local Organizing Committee of the SEAC 2001 conference and personally Dr. Mary Blomberg and Dr. Göran Henriksson for great help in providing the possibility to present this paper to the scientific community. I would like to thank as well Ms. Rita Balkute (Archives of people's art, Lithuanian Folk Culture Centre) and Mr. Rytis Ambrazevičius (Department of Ethnomusicology, Lithuanian Academy of Music) for permission to use their photos in this paper. I am grateful to Mr. Saulius Lovčikas (Department of Astrophysics, Institute of Theoretical Physics and Astronomy) and Mr. Dainius Razauskas (Lithuanian Folk Culture Centre) for help in preparing the English version of this paper. I am most grateful to Dr. Arnold Lebeuf (Institute Religioznawstwa, Jagielonian University) for reading and substantially improving the manuscript.

Persons providing information

1) *Balevičiūtė Jadvyga*, born 1925, Balkasodis village, Miroslavo district, Alytus region. Written down by V. Vaitkevičius, 1993.
2) *Baulienė Ona*, born 1919, Žiežmariai village, Kaišiadorys region. Written down by A. Morkūnaitė, 1994.
3) *Maskeliūnas Liudas*, born 1907, Lapšiai village, Seirijai district, Lazdijai region. Written down by A. Vaicekauskas, 1984.
4) *Mikalauskaitė Salomeja,* born 1924, Griškabūdis village, Griškabūdis district, Šakiai region. Written down by E. Žiemys, 1992.
5) *Vilkevičius Juozas,* born 1909, Paluobiai village, Griškabūdis district, Šakiai region. Written down by J. Šorys, 1992.
6) *Žakauskas Bronius*, born 1912, Padumbliai village, Kapčiamiestis district, Lazdijai region. Written down by A. Morkūnaitė, 1992.

*Institute of Lithuanian Scientific Society, Museum of Molėtai District, P.D. 51, Molėtai, LT-415, Lithuanian, j.vaiskunas@mail.lt.

References

Balys, J. (compiler). 1935-1940. The *Synopsis of Beliefs*. Compiled from the Archives of Lithuanian folklore (LTA). This manuscript now belongs to The Department of Ethnology of the Institute of History at the Lithuanian Academy of Sciences. Vilnius.

Balys, J. 1993. *Lietuvių kalendorinės šventės*. Vilnius.

Daukantas, S. 1976. *Raštai*, vol. 1. Vilnius.

Kudirka, J, 1992. *Užgavėnės*. Vilnius.

Lovčikas, S. 1996. Review of archaeoastronomical objects in Lithuania. In P*roceedings of the Second SEAC Conference, Bochum, August 29th-31st, 1994*, ed. W. Schlosser, 139-149. Bochum.

Stryjkowski, M. 1846. *Kronika Polska, Litewska, Żmudyka i wsyzstkej Rusi. Wydanie nowe dokładnim powtórzeniem wydania pierwotnego królewieckiego z roku 1582*, vol. 1. Warsaw.

Thurston, H. 1994. *Early Astronomy*. New York.

Trinkūnas, J. 2000. *Baltų tikėjimas*. Vilnius.

Measuring time in the Central Rhodopes

Vesselina Koleva*

Abstract
The present paper presents a study of the time-measuring practices of Orthodox and Muslim Bulgarians in the central Rhodope Mountains (Rhodopes) according to ethnographic and folklore data from the 19th and 20th centuries. The basic measurement units and methods are discussed. Included is an examination of the so-called 'shepherd year', in use until the 19th century. The author concludes that alongside official time-measuring practices the local population has preserved numerous ancient traditions.

Introduction
The Rhodopes are situated in southern Bulgaria. Their location has determined the peculiar fate of the population during different historical periods and the merging of various cultures. Thracians, Slavs, Proto-Bulgarians, Romans, Greeks, Turks, and others have left their traces in the time-measuring traditions and calendar practices in the region (V. Koleva 1999: 195-197). The Rhodopes were incorporated within the Bulgarian state in the 9th century, when the conversion to Orthodox Christianity also took place. Since the end of the 14th until the end of the 19th century, Bulgaria was part of the Ottoman Empire whose assimilation policy was especially strong in the Rhodopes. The religion and language of the local population were strongly influenced, and the Turkic sheep breeders, called *yurùk* (from Turkish *yürük*), who were moved there from Asia Minor, turned their occupation into the main means of living in the area.

The present research is based on ethnographic and folklore data collected in the 19th and 20th centuries and comprises information about folk astronomy and time-measuring practices in the central Rhodopes. Data from questionnaires and personal observations of the author have also been used.

Division of time
Day and night The basic division of time into *den*, day, and *nosht*, night, reflects the visible movement of the sun. Sunrise, sunset, and upper and lower culminations of the sun divide this day-night period into four parts. In the central Rhodopes these moments are called *ìznik*, rise (sunrise), *zàni*, decline (sunset), *plàdne*, noon, and *sredènosht*, middle of the night (midnight). These are also the names for the four cardinal points! The differences in sky light define smaller parts of one and a half to three hours in the day and night, which are also used for religious purposes. People use more than 30 terms to refer to different moments and time intervals during the day and night (Shishkov 1965: 360-362). Some of them are clearly related to the twilight, e.g. the astronomical twilight in the morning is referred to in the expressions *zàbela*, it became lighter, *sabàh karshì* , before daybreak (from Turkish *sabah karşi*,) and the civil twilight is presented as *pùkna se zorà*, day broke. In the mountains, people distinguish between the moments when the sun lights up the top of the peaks, *ìzgref slònce*, sunrise, and later their lower slopes, *nàgref slònce*, the warmth of the sun. Prayer time is referred to by specific terms: *klepàlno vrème*, time to ring the bells (two hours before dusk when the bells chime in Christian churches), or *ikindìya*, the middle of the period between noon and sunset (the time for the afternoon prayer of the Muslims, from Turkish *ikindi*,), and *yaciya*, two hours after sunset (the beginning of the night and the time for their night prayer, from Turkish *yatsi*). Morning breakfast takes place at *golèma prògima*, big breakfast (from Greek *πρόγευμα*), and at *plàdne* people have their lunch. Some two hours after dusk the so-called *golèma vechèrya*, big dinner, takes place and after it the mysterious time of the evil spirits, *spotàyno vrème*, lurking time, sets in. After midnight, at *petlèno vrème*, cockcrow, those evil spirits are chased away.

Week The folk name for week is *nedèlya* and a rarer name is *sèdmica* (from *sèdem*, seven). Initially *nedèlya*, don't-work day (Sunday), was the name of the first day of the week and presently it refers to its last day. The formerly last day, the seventh day of the week, has a similar meaning, *sàbota* (Saturday), which comes from the ancient Babylonian *Shabbat* (peace, calm). The second day of the week was *ponedèlnik*, after don't-work day (Monday). The third was *vtòrnik*, which derives from the word 'second', but should be interpreted as the second day after Sunday. The next day, *sryàda*, means 'the middle'. It is followed by *chetvàrtak* meaning 'fourth day" and *pètak* meaning 'fifth day'.

Natural periods, lasting about one week, are defined by the main phases of the moon. The first three moon phases are called *mlàda mèsechina*, young moon (waxing half moon), *pòlna mèsechina*, full moon, and *ùsip* (in Russian *усыпить* means 'to put to sleep, to kill'), or *kràtesh*, ending, dying (waning half moon). In a moonless period, the moon is *kràtena*, gone or dead, and after that it *se ràzhda*, is born, or it is *novena*, new (new moon) (Boneva 1994: 11).

Months and years are defined according to the official calendars: the Julian solar calendar of the Christians (V. Koleva 1999: 196) and the lunar and solar calendars of the Muslims. The Rhodopean Christians (a) and Muslims (b) call the months (*mèsec*) from January (1) to December (12) by the following names: 1a *Golemìn*, 1b *Golèmet mèsec*; 2a *Malkìn*, 2b *Màlket mèsec*; 3a *Màrta* , 3b *Màrta*; 4a *Lòshko*, 4b *Lòshko*; 5a *Yòchko*, 5b *Gèrgyovden*; 6a *Tòduroske*, 6b *Prèdoy*; 7a *Pètruske*, 7b *Pètruvden*; 8a *Yègus*, 8b *Yègus*; 9a *Rùyen*, 9b *Golèmata chòrkva*; 10a *Dimìtroske*, 10b *Màlkata chòrkva*; 11a *Gergyòske*, 11b *Kàsum*; 12a *Nikùltske*, 12b *Kòlada* or *Bozhìch* (Zaimov 1954: 106-141; Raychevski 1998: 91-92).

Major Christian festivals have given the names 5b, 6a, 7a, b, 9b, 10a, b, 11a, b, and 12a, b. The names of months 9b, 10b, and 12b are, in fact, the older names for the festivals *Golèma Bogoròdica* (15 August), *Màlka Bogoròdica* (8 September), and *Kòlada* (25 December). These old names were preserved by the Bulgarians who adopted the Muslim religion. *Prèdoy* (6b), first milking, is the name of a major festival in the shepherds' economic calendar, which coincided with the day of St. Constantin and St. Elena (21 May). In most cases these festivals are in the respective preceding months. This discrepancy can be explained with the inaccuracy of the Julian calendar, used officially in Bulgaria until 1916, or with the old lunisolar calendar tradition (V. Koleva 1999: 195-196). The ancient Slavic lunisolar calendar has given the name 9a and the names 4a, b, and 5a, which are dialect forms of the Old Slavic month names *lazhitràv* and *izòk* respectively (Zaimov 1954: 119, 122, 131). Month names used in northern Greece, *Μεγάλος μήνας* and *Μικρός μήνας*, have been borrowed for 1a, b and 2a, b (Zaimov 1954: 109, 115). The name *Gergyòske* (11a) is more specific and is connected with the festival *Zìmen Svetì Geòrgi*, Winter St. George (Zaimov 1954; 139). The name *Kàsum* (11b), which is a name of the festival *Dimìtrovden* (26 October), comes from the Turkish solar calendar where *Kàsim* is the name both of November and of the date 7 November. The Arabic word *kasim* means 'something that divides'. The second date, which is used to divide the year into two, is 23 April (the festival *Gergyòvden*). The Rhodopean Muslims call it *Hèdreles*. In the Turkish calendar this festival is called *Hidralez* and is on 6 May.

Seasons An earlier division into seasons was that of *zìma* (winter) and *lèto* (summer), and later the transitional seasons of *pròlete* (pre-summer or spring) and *pòdzime* (pre-winter or autumn) were defined. But the traditional division into a warm and a cold period has remained in use. The limits of these periods are *Gergyòvden* (23 April) and *Dimìtrovden* (26 October). On Rhodopean wooden calendars (*rabosh*) from the 18th and 19th centuries the days are marked only on *two* of the four edges of the stick, and the starting dates are 26 October and 23 April, 1 November and 1 May, or 1 September and 1 March (E. Koleva 1971: 247-270). On a more recent, unpublished *rabosh* the starting date is 1 January, and the days are carved in three-month periods on the *four* edges of the stick.

The warming of the weather begins on *Sveti chetirsi*, the Forty Saints (the Holy Great Forty Martyrs) or *kàrkchibuk*, forty chibouks (from Turkish *kirk* and *çubuk*), which is 9 March when "40 cold needles are pulled out and 40 hot needles are stuck into the earth" (Raychevski 1998: 29). The Julian date, corrected by 12 days, coincides with the spring equinox (21 March). People also know that the sun 'turns around' on *Svetì Vartolomèy* (11 June), and from *Svetì Elisèy* (14 June) till *Lèten Atanàsovden* (5 July) it 'walks back' to the winter (Boneva 1994: 10). On *Ènyovden* (24 June) at sunrise people observe how the sun 'rolls over', 'plays' (Raychevski 1998: 52). After *Ilìnden* (20 July), *sredè lèto*, midsummer, people already say that the summer is going away. Autumn sets in on *Golèma Bogoròdica* (15 August) (Shishkov 1965: 363). The weather turns cool and the snow comes after *Dimìtrovden* (26 October), and the real winter lasts from *Arhàngelovden* (8 November) till *Atanàsovden* (18 January), which is called *sredè zìma*, midwinter.

Evidence that people knew about the movement of the sun can also be found in ethnochoreographic data. The oldest layer of the Bulgarian dance tradition has been preserved among the Muslims in the Rhodopes (Tyankova 1994: 267). They dance during the *Bayràm* festivals, each Sunday and Friday, on *Hèdreles*, on *Prèdoy*, on *Kàsum*, and on *Kòlada*. The nature of the dances depends on the season and the part of the day. In wintertime and at night, dances are livelier, with more leaps and changes of direction, while the spring dances are soft, 'walking' dances, and games are played with quiet and calm songs. During the day, dances are calmer and open, while at sunset their steps become smaller and quicker and they turn into a spiral (Tyankova 1994: 281). Therefore, in the language of the 'magic' dances the visible movement of the sun is depicted: with the quickstep, leaps, and change of direction around the winter solstice, and at night the sun moves towards the summer and the sunrise respectively, while in the spring and daytime its steps are calm and broad.

Measuring Time

The equally long hours of the 'European' and 'Turkish' systems are well known, but their use is not common in everyday life (*Letostruy* for 1869: 13). The hour (*chas, sahàt,* from Turkish *saat*) is not always 1/24 of the night-and-day period. Even the well-to-do Muslims, who possessed watches, did not have an exact notion of one-hour's length of time. In the 'Turkish way' they count 2 x 12 *equal* hours *from sunset till sunset* and periodically set their watches. Traditionally, time is fixed by the length and direction of a shadow cast by a vertical object (gnomon); thus the 'hour' depends on the length of the day during the year. Such a 'watch' can be found in a medieval literary source (Kristanov and Duychev 1954: 373-374) where the 12 daytime hours in each month are represented in a certain number of steps. The first hour of December, at sunrise, is the longest—33 steps, and the shortest is the sixth hour of June at noon—two steps. People generally know that at noon a man's shadow is shortest and it is 'a step and a half or two steps' long, but they never refer to a particular season. The sun's altitude can be measured by a 'man's height' (Shishkov 1965: 360), by comparing the shadow cast by a rock or a tree with that of a person. As a whole, anthropometrical units are widely used in the Rhodopes. For example, angles are compared with the length of the forefinger, which is one *rùp* long or ca 8 cm (one *rùp* is also the width of the four fingers held together, with thumb excluded). When measuring with a stretched-out arm, one *rup* corresponds to an angle of ca 9°, and the distance between the spread thumb and forefinger, *chèparek*, is ca 18°. Small time intervals of one minute to one hour are defined in expressions related to the length of the day around the winter solstice and to the sun's barely visible movement away from its southernmost points of rising or setting. Such metaphorical expressions (which are sometimes inaccurate) featuring 'linear' units are: the day grows 'by a needle's point' as of *Varvàra* (4 December), or 'by a millet' as of *Andrèevden* (30 November), or 'by a cock's leap' from *Andrèevden* till *Vasìlyovden* (1 January), or 'by a three-year-old deer's leap' till *Atanàsovden* (18 January) (Popov 1991: 35). In the daytime, the sun's position with respect to the mountains plays the role of a watch and a seasonal indicator. In the night such benchmarks are the well-known luminaries Venus, Jupiter, Arcturus, or Sirius ('the shepherd's star'), as well as the constellations Big Dipper (*Kolàta*, cart), Orion (Oràlitsa, plough), and the star cluster Pleiades (*Kvàchka*, brood-hen) (Kovachev 1914: 22-23; Boneva 1994: 13).

The Calendar

Along with the official calendars the Rhodopean shepherds widely use an economic calendar, the so-called 'shepherd year' (Dechov 1968: 329-330). The 'shepherd year' is divided into two parts by the dates 26 October and 23 April, when the old labour contracts expire and the new ones are concluded. *Dimitrovden* is the starting, 'zero' day. The counting begins from 27 October and ends on 25 October. The numbers in the two periods are counted out in the *Turkish language* from 1 to 120 without a break and, after that, the rest of the days till the end of each period are counted backwards. Months and weeks are not mentioned. Economic periods are expressed in days: the insemination of the sheep begins on *irmi beshe*, 25 days before *Dimìtrovden* (Dechov 1968: 299); the forming of new herds with pregnant sheep starts from *doksan*, 90 days after *Dimìtrovden*; sheep begin to lamb after *yuz irmi*, 120 days after *Dimìtrovden* (Dechov 1968: 252). Thus the first half of the year has 179=120+59 days, and the second half has 186=120+(65+1) days; i.e. the year is divided into four parts in a ratio of ca. 2:1:2:1. This suggests a possible division into six parts with five to six remaining days: 60+60+59 (+1 day in a leap year) and 60 60+60+(5+1). Here, dividing dates become 26 October, 25 December, 23 February, 23 April, 22 June, 21 August, and 20 October.

Most of these dates used to be seasonal borders *during different epochs*. A next natural step in the division of the year could be the division into 30-day periods and five epagomenal days, typical of the movable years in the ancient Egyptian and Old Persian calendars. Similar backward day counting, as in the Rhodopean 'shepherd' calendar, was found in Athens in Solon's time (ca 640-560 BC), where the days in the third decade of the lunar month were counted backwards. A well-known example of backward day counting is the enumerating of the days in the Roman calendar as so many before each of the three basic times of the month: the *calends*, the *nones*, and the *ides*.

Conclusions

The isolation of the population in the central Rhodopes has resulted in the preservation of ancient traditions that are intricately interwoven and very hard to analyse. The Rhodopean time-measuring practices and folk calendar have counterparts in neighbouring Aegean Thrace and in Asia Minor. Most probably, however, they descend from the ancient civilisations of Egypt and Babylon.

National Astronomical Observatory, P.O. Box 135, 4700 Smolyan, Bulgaria,
vpk_smolian@yahoo.com.

Bibliography

Boneva, T. 1994. People's world outlook (Бонева, Т. Народен светоглед, в *Родопи. Тадиционна народна духовна и социално нормативна култура*, ред. Р. Попов и С. Гребенарова. София, 7-50).

Dechov, V. 1968. Sheep breeding in the Central Rhodopes (Дечов, Васил Среднородопско овчарство, в *Избрани произведения*. Пловдив, 197-341).

Koleva, E. 1971. Folk calendars - Raboshi (Колева, Е. Народни календари-рабоши, в *Известия на българските музеи*, vol. 1, 247-270).

Koleva, V. 1999. The Calendar in Medieval Bulgaria. In *Oxford VI and SEAC 99. Astronomy and cultural diversity*, ed. C. Esteban and J. A. Belmonte, 195-201. Santa Cruz de Tenerife.

Kovachev, Y. 1914. Folk astronomy and meteorology (Ковачев, Й. Народна астрономия и метеорология. *Сборник за народни умотворения и народопис*, vol. 30, София, 1-85).

Kristanov, Ts. and Duychev, I. 1954. *Natural-scientific knowledge in Medieval Bulgaria: Miscellany of historical sources* (Кристанов, Цв. и Дуйчев, Ив). *Естествознанието в средновековна България. Сборник от исторически извори.* София).

Letostruy or Home calendar for the common year 1869 (*Летоструй или домашен календар за проста година.* Пловдив).

Popov, R. 1991. *Saints twins in the Bulgarian folk calendar* (Попов, Рачко. *Светци близнаци в българския народен календар.* София).

Raychevski, S. 1998. *Rhodopean folk calendar* (Райчевски, Стоян. *Родопски народен календар.* София).

Shishkov, S. 1965. Division and measuring of time in the Rhodopes (Шишков, Стою. Разпределение и пресмятане на времето в Родопите, в *Избрани произведения*. Пловдив, 358-363).

Tyankova, Y. 1994. Games and dances (Тянкова, Й. Игри и танци, в *Родопи. Традиционна народна духовна и социално нормативна култура*, ред. Р. Попов и С. Гребенарова. София, 267-284).

Zaimov, J. 1954. Bulgarian folk names of months (Заимов, Й. Български народни имена на месеците, в *Известия на Института за български език,* vol. 3, 101-147).

Archaeoastronomical research on the Kurgans 'with moustaches' in the south Trans-Urals: Results from a preliminary study of the calendar systems and world outlook of the nomads of the first millennium AD

Dmitry G. Zdanovich* and Andrey K. Kirillov**

Abstract

The south Trans-Urals is one of two regions with a high concentration of the so-called kurgans 'with moustaches', the other being central Kazakhstan. It was noted long ago that their structure is related to the azimuths of main events of the yearly solar cycle. For the first time the materials on the archaeoastronomy of these kurgans are presented, which differ from those in central Kazakhstan by a series of peculiarities. The authors present the general directions of the archaeoastronomical research and introduce some new directions. The main issues of this research are the following: astronomical and calendar functions of the kurgans 'with moustaches', their orientation in the landscape with respect to astronomically significant directions, the dating of these kurgans using archaeoastronomical methods, and the role of angular values connected with the 'golden section' in the design of these sites. The research was conducted in the field with a theodolite and also in the laboratory. A computer modeling of azimuths of main astronomical events was used. A more or less complete study of nine kurgans 'with moustaches' has been performed.

The basic results of our research allow us to make the following conclusions: The kurgans 'with moustaches' could play the role of burial kurgans, although they frequently had some special independent function related to solar cults. A typical kurgan 'with moustaches' was not an instrument for keeping a calendar. It was rather a ritual object related to calendar symbols. Nevertheless, the data from the kurgans 'with moustaches' permits the definition of two different approaches to counting calendar time on the basis of the yearly solar cycle. The design of some of the kurgans contains the azimuths of 'high' and 'low' moon setting and rising in extreme positions and angular values that conform to the definition of a 'golden section'. The absolute dating of these kurgans is a separate problem. The dates obtained by archaeoastronomical methods point to the first millennium AD as the time for their construction. The kurgans 'with moustaches' reflect the development of astronomical knowledge and the growth of the tradition of astronomical observations in the Urals-Kazakhstan region in ancient times.

In prehistoric Europe the study of the motions of the sun and the moon was closely connected to the symbolism of life and death. The results of astronomic observations were materialized first of all in ritual architecture (Ruggles 1996: 15-27). Through millennia kurgans were a dominating form of cult architecture in the south Trans-Urals. The 'kurgan tradition' forms here no later than the third millennium BC and keeps developing up to Medieval times. Different Indo-European groups, including Iranians, Ugric, and Turkic peoples, were the bearers of this tradition. Currently, a complex interdisciplinary approach to the research of kurgan objects is under development. Within this approach different authors have studied the astronomical aspects of several kurgans in Russia and Kazakhstan. Methods of archaeoastronomy have been used for reconstructing aspects of the world outlook of kurgan burials (Marsadolov 2000: 30) and for testing the hypothesis of calendar functions of some forms of kurgan architecture, the so-called kurgans 'with moustaches' (Bekbasarov 1997: 11-13; 1998: 163-170; Beisenov 1998: 171-175).

The present work continues these directions in research and also contains some new approaches to the study of the archaeoastronomy of kurgans. These relate in particular to dating kurgan objects using archaeoastronomical methods. The main attention is paid to the kurgans 'with moustaches', which we refer to early Medieval times.

The south Urals is the region with the second highest concentration of the kurgans 'with moustaches', the highest being in central Kazakhstan (Kukushkin 1993; Lyubchansky 1998; Lyubchansky and Tairova 1999). The publication of the extensive material from these kurgans, with their cartography and typology, was first done for central Kazakhstan (Margulan et al. 1966: 305-311). They were related to the Tasmolinskaya culture of the Early Iron Age and dated to the 7th-3rd centuries

BC. Many Kazakhstan researchers still hold this opinion (Beisenov 1996: 113). Currently, however, the dating and cultural characteristics of these sites are under revision. In the opinion of Lyubchansky the kurgans 'with moustaches' should be dated to the 5th-8th centuries AD. They were left behind by groups of Iranian and Turkic-speaking groups of post-Hun times (Lyubchansky 1998: 303-310). According to S. G. Botalov the kurgans 'with moustaches are early Turkic sites of the 6th-9th centuries AD (Botalov 1998: 321-330).

The authors present for the first time the results of archaeoastronomcal studies of the kurgans 'with moustaches' of the south Urals, which differ from the Kazakhstan ones by a series of peculiarities. A more or less complete study of nine kurgans, with measurements made in the field with a theodolite, was performed over the period from 1993 to 2000 (Fig. 1, Table 1). In order to model the sequence of observations of the main astronomical events, the computer program *Turbosky* was used.

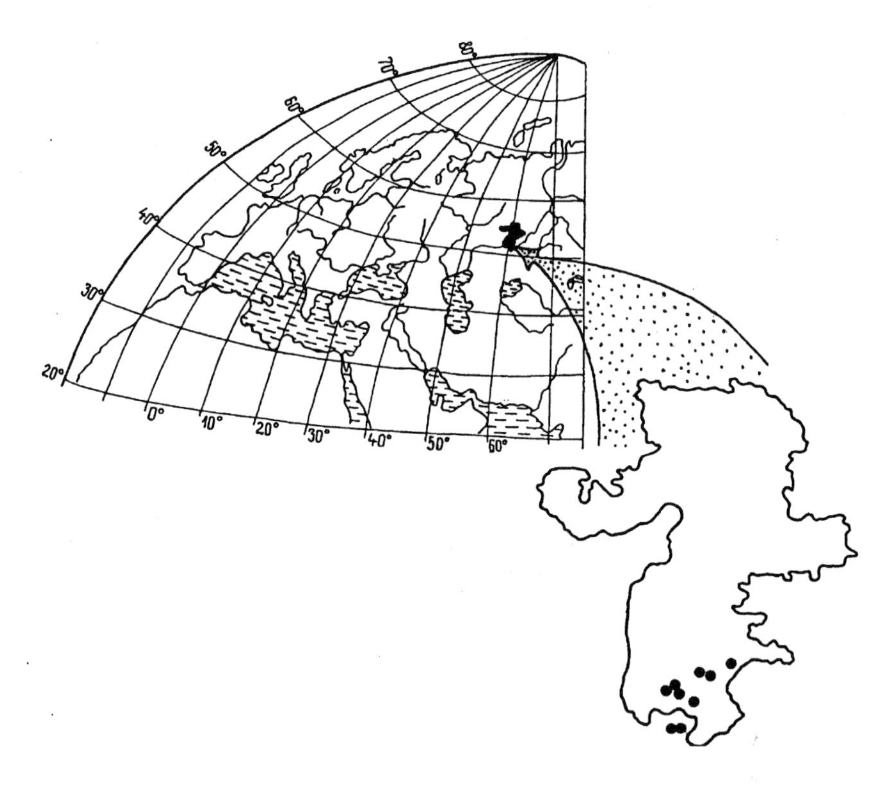

Fig. 1. Disposition of the investigated kurgans 'with moustaches' on a map of the Chelyabinsk region, south Trans-Urals, Russia.

Site	Latitude φ	Length of moustaches (m)		Values of basic angle of structure	Azimuth of the axis	Dates in years AD	
		Northern	Southern				
Gorodischenskoe IX	52°59'	234	235	49.00°	87° 23'	—	500-700
Cherkasinsky	52°41'	137	134	49.33°	95°40'	—	—
Solonchanka I	52°12'	204	200	40.38°	98°30'	—	400-525
Solonchanka IX	52°11'	91	111	52.67°	105°41'	1015	ca 400
Rymniksky	52°30'	155	147	47.03°	?	—	—
Kondurovsky	52°37'	242	185	44.10°	108°35'	460	—
Amursky	52°37'	112	112	45.80°	104°20'	—	—
Kamenny Ambar VII	52°48'	110	100	52.20°	108°57'	970	—
Elizavetpolsky	52°50'	47	39	41.97°	111°45'	350	400-500

Table 1. Summary of the data on the kurgans 'with moustaches'. In the 'Dates' column those to the right are the results of archaeological excavation and those to the left have been reached through archaeoastronomical methods.

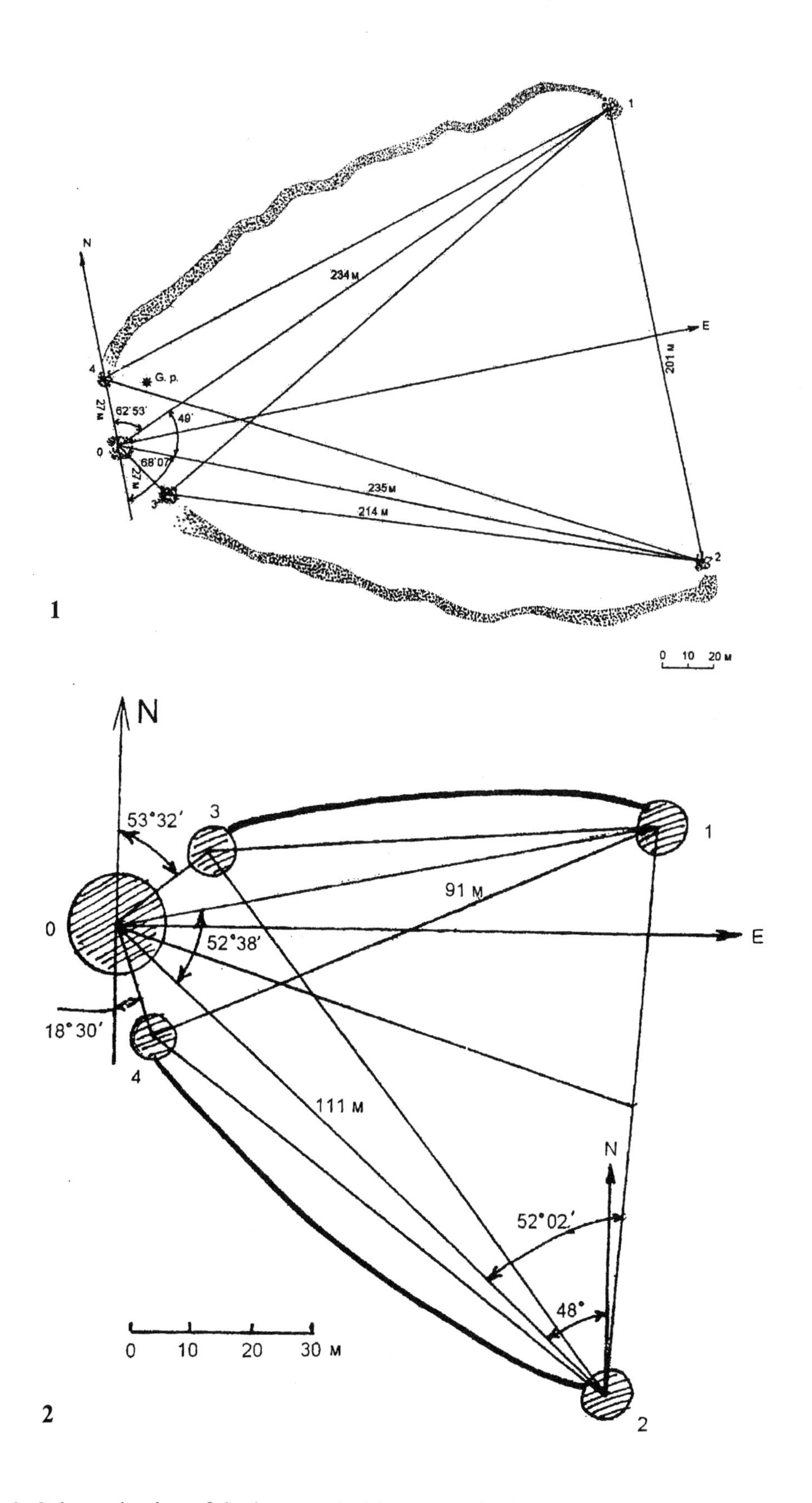

Fig. 2. Schematic plan of the kurgans 'with moustaches': 1) Gorodischenskoe IX; 2) Solonchanka IX.

The research performed allows us to draw the following basic conclusions:

1) The kurgans 'with moustaches' could play the role of burial kurgans, but more often they had some special independent functions. The architecture of these kurgans does not have vivid material

analogies with the preceding ages; undoubtedly it had developed in a non-material sphere of myth and ritual. The attempt to reconstruct this sphere using the methods of archaeoastronomy returns us to old ideas of certain archaeologists (Margulan et al. 1966) concerning the relationship of the kurgans 'with moustaches' to solar cults. It was noted long ago that the 'moustaches' of such kurgans are oriented to the east (Margulan et al. 1966: 310); a more detailed study permitted their division into two groups. The first one is represented by the kurgans where the axis of symmetry of the 'moustaches' (the main axis of the structure) almost completely coincides with the direction to the east (Fig. 2:1). The kurgans of the second group have this axis deviating to the south (Fig. 2:2). The value of the deviation lies within the range of 14°-22° (Table 1).

2) The study of the kurgans 'with moustaches' in Kazakhstan (Bekbasarov 1997: 11-13; 1998: 163-170) and in the Trans-Urals shows that their design reflects the azimuths of main events of the yearly solar cycle. In our opinion a typical kurgan 'with moustaches' was not a good instrument for keeping a calendar. It was rather a ritual object related to calendar symbols. Nevertheless, the data from these kurgans allow us to identify two different approaches to counting calendar time. In one calculation system (a group of kurgans with east-west orientation of the axis: the Gorodischenskoe IX and Cherkasinsky sites), the starting point was the day (days) of the equinox, and the main task evidently was to fix the time intervals of certain durations adjacent to these days. A good example comes from Gorodischenskoe IX (Fig. 2:1). On the condition that the observer was in the center of the main kurgan of the site, it was possible to observe sunrise twice a year above the ends of each 'moustache':

0→1, a = 62°53', 22 April and 16 August.
0→2, a = 111°53', 12 February and 22 October.

The intervals could have had a significant religious and social meaning. For example, in the case of the mountain Tadjiks, all community works of the inhabitants of the kishlak were timed to the period 'chilla', i.e. 40 days after the winter solstice (Maysky 1934: 103-107). The calculation system of the other group (kurgans with an axis deviating to the south: Kondurovsky, Amursky, Solonchanka IX, etc.) is related to calendar time-counting proper; the starting points are the days of the solstices; the counting could go on all the year long (Fig. 2:2; Table 2).

We cannot exclude the possibility that these two groups of kurgans 'with moustaches' are chronologically different one from another. They may also reflect different cultural and ethnic traditions. In addition, the specifics of traditional perceptions of the category 'time' and ways of measuring it allow the existence of several calendar systems within one tradition (Zdanovich 1998: 147-161)

Direction	Geodetic azimuth	Phenomenon	Date according to the modern calendar
2→0	47°59'	setting	22 June, summer solstice
2→4	53°04'	setting	11 July
4→1	67°03'	rising	10 August
0→1	79°22'	rising	1 September
3→1	87°16'	rising	13 September
1→0	100°38'	setting	3 October
1→4	112°57'	setting	25 October
4→2	126°56'	rising	22 November
0→1	133°01'	rising	22 December, winter solstice
4→2	126°56'	rising	10 January
1→4	112°57'	setting	6 February
1→0	100°38'	setting	26 February
1→3	92°44'	setting	10 March
3→1	87°16'	rising	21 March, equinox
0→1	79°22'	rising	1 May
4→1	53°04'	setting	24 May
2→0	47°59'	setting	22 June, summer solstice

Table 2. Sequence of observations of the sun from the kurgan 'with moustaches' Solonchanka IX. See Fig. 2:2 for the directions.

3) The design of some of the kurgans 'with moustaches' contains the azimuths of 'high' and 'low' moon setting and rising in extreme positions. Unlike the solar azimuths these directions are not systematic, and it is hard to draw certain conclusions.
4) The design of these kurgans also contains angular values that conform to the definition of a 'golden section'. Sometimes, e.g. in the case of the Amursky and Cherkasinsky kurgans, such values characterize the orientation of the objects to the dominating points of the surrounding landscape. The problem of proportionality of the parts of the kurgans 'with moustaches' and the standards of 'beauty' reflected in their design requires more research.
5) The absolute dating of the kurgans 'with moustaches' is a separate problem, and the discussion is still going on. It is also relevant for archaeologists. Dating such objects by archaeoastronomical methods is problematic in general. The Solonchanka IX kurgan exemplifies the problem. The epoch of its construction, defined by the azimuth of the upper limb of the sun at summer solstice sunset and of the moon setting at the northern major standstill, corresponds to 1015 AD. If, instead, the moment of setting is accepted as the horizon contact by the lower limbs of the sun and moon, then the hypothetical date will correspond approximately to 3500 BC. The dates obtained by archaeoastronomical methods at least allow us to point to the epoch of the construction of the kurgans, the first millennium AD (Table 1). The archaeoastronomical date for the Kondurovsky kurgan 'with moustaches' obtained by the lower limb of the solar disc, 460 AD, looks quite trustworthy from the point of view of modern archaeology. Thus the dates obtained allow us to accept the possibility of dating the kurgans 'with moustaches' of the south Trans-Urals to the 5th-8th centuries AD (Lyubchansky 1998: 303-310). We can assume that the tendency to date the Kazakhstan kurgans of the same type to the 7th-3th centuries BC is related to the existence of a different chronology, according to which the earlier, mainly burial kurgans are later supplemented by kurgans 'with moustaches'.
6) In constructing kurgans 'with moustaches', the Iranian and Turkic-speaking groups of the 5th-8th centuries AD had probably wanted to demonstrate their participation in the preceding tradition. The fact of the orientation of these kurgans to astronomically significant azimuths and their relationships to the ruins of the fortified settlements of the Bronze Age (Arkaim, Alandskoe) as well as to the dominating points of the landscape in the south Trans-Urals evidently represent the same ideological aspect.

Being the main form of cult architecture of the region for thousands of years, the kurgan sites of the south Trans-Urals are organically related to the practice and ideology of ancient astronomy. The millennial cultural tradition, including the kurgans 'with moustaches', reflects the accumulation of astronomical knowledge and the stable development of methods of astronomical observations. The origins of this ancient astronomical tradition in the Urals-Kazakhstan region probably go back to the pre-kurgan age.

*Chelyabinsk State University, Institute of History and Archaeology of the Urals Department of Russian Academy of Sciences, Chelyabinsk, Russia, root@arcaim.cgu.chel.su.
**Chelyabinsk State Pedagogical University, Chelyabinsk, Russia, kirillov@cspi.urc.ac.ru.

References

Beisenov, A. Z. 1996. Nekotorye voprosy izucheniya kultovo-ritualnykh kurganov s kamennymi gryadami Kazakhstana na sovremennom etape, in *Zhrechestvo i shamanizm v skifskuyu epokhu*, ed. A. Yu. Alexeev et al., 112-122. St. Petersburg.

Beisenov, A. Z. 1998. Ob astronomicheskikh issledovaniyakh na kurganakh s kamennymi gryadami. In *Voprosy arkheologii Kazakhstana* 2, ed. Z. Samashev. Moscow.

Bekbasarov, N. M. 1997. Arkheoastronomicheskie issledovaniya atasuskikh kurganov "s usami". In *Drevnyaya astronomiya: nebo i chelovek*, ed. V. V. Volkov et al., 11-13. Moscow.

Bekbasarov, N. M. 1998. Kurgan 's usami' na reke Saga: astronomicheskoe soderzhanie raspolozheniya ego elementov. In *Voprasy arkheologii Kazakhstana* 2, ed. Z. Samashev, 163-170. Moscow.

Botalov, S. G. 1998. Ranneturkskie pamyatniki Uralo-Kazakhstanskikh stepei. In *Kultury evraziiskikh stepei vtoroi poloviny I tysyacheletiya n.e: Voprosy khronologii*, ed. D. A. Stashenkov, 321-330. Samara.

Kukushkin, I. A. 1993. Simvolika solyarnykh kultov v megaliticheskikh pamyatnikakh Kazakhstana. In *Kochevniki uralo-kazakhstanskikh stepei*, ed. A. D.Tairov, 68-71. Ekaterinburg.

Lyubchansky, I. E. 1998. Khronologicheskie aspekty komplexov kurganov "s usami" evraziiskoi stepi. In *Kultury evraziiskikh stepei vtoroi poloviny I tysyacheletiya n.: Voprosy khronologii*, ed. D. A. Stashenkov, 303-310. Samara.

Lyubchansky, I. E. and Tairova, D. 1999. Arkheologicheskoe issledovanie kompleksa kurgan 's usami' Solonchanks I. In *Kurgan 's usami' Solonchanka I.* Works of the Museum Arkaim, ed. A. D. Tairov, 5-62. Ekaterinburg.

Margulan, Alkei Kh. et al. 1966. *Drevnyaya istoriya Tzentralnogo Khazaknstana*. Alma-Ata.

Marsadolov, L. S. 2000. *Arkheologicheskie pamyatniki IX-III vv. do n.e. gornykh raionov Altaya kak kulturno-istoricheskii istochnik (phenomen pazyrykskoi kultury)*. St. Petersburg.

Maysky, L. 1934. Ischislenie polevogo perioda selskokhozaistvennykh rabot u gortzev Pamira i Verkhnego Vancha. *Sovetskaya Etnografia* **4**, 102-107.

Ruggles, C. 1996. Archaeoastronomy in Europe. In *Astronomy before the telescope*, ed. C. Walker, 15-27. London.

Zdanovich, D. G. 1998. Mifologicheskoe vremya i ego ischislenie (po materialam ugro-samodiiskoi etnografii). In *Voprosy arkheologii Urala* 23, ed. V. T.Kovalyova, 147-161. Ekaterinburg.

The anthropoid in the sky: Does a 32,000-year old ivory plate show the constellation Orion combined with a pregnancy calendar?

Michael Rappenglück*

Abstract
A very small ivory plate excavated in the cave of Geißenklösterle, Germany, shows a manlike being in low relief on one side and rows of notches on the other side and along the edges. Based on archaeological data, the phenomenological description of the plate, the astronomical reconstruction of the sky in the Aurignacian epoch, and ethnoastronomical arguments, the research work establishes the hypothesis that the anthropoid represents Orion at the vernal equinox ca 30,000 BC, combined with a special type of pregnancy calendar related to the heliacal rising and setting of Betelgeuse.

One of the oldest representations of an anthropoid figure worldwide was unearthed in the Geißenklösterle cave near Blaubeuren, Germany (Φ: 48° 23.9' N | λ: 9° 46.5' E, 585 m above mean sea level, lower level II b). On one side of the small rectangular plate there is in low relief a manlike being in adorant style (Fig. 1A). It may be partly formed of a human and a feline shown at the pounce (Hahn 1986: 119). Between the legs there is a long artificial extension on the body's axis, which goes down to the level of the right heel. On the other side (Fig. 1B) and at all four edges (Fig. 1A: a, b, c, d), rows of different kinds of notches, made intentionally and periodically, are discernible (Hahn 1986: 118-119; Marshack 1987: 143; Bosinski 1994: 77). At the edges 39 cuts can be counted: 6 (a), 13 (b), 7 (c), and 13 (d). On side B the notches are arranged into four columns: 13 (v_1), 10 (v_2), 12 (v_3), and 13 (v_4), summing up to 48 (Hahn 1986: 118-119; Müller Beck et al. 2001: 65). As a few researchers count ± 1 notch in columns v_2 or v_3, the total number of notches could be 48 ± 1 (Bosinski 1982: 77; Marshack 1987: 143).

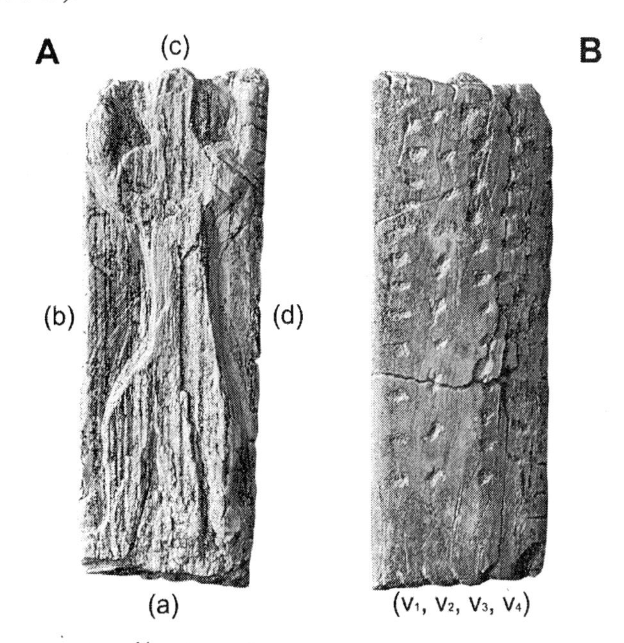

Fig. 1. The small mammoth-ivory plate, 38 x 14, 1 x 4,5 mm. (Württembergisches Landesmuseum Stuttgart, Inv. No. S 89, 14). Sides A and B. Photo: Bosinski 1990: 69; courtesy Gerhard Bosinski.

There exist five ^{14}C dates for the lower level II b (Djindjian et al. 1999: 180, 165-166, 172, 404-405): 33,700 ± 825 BP (H 4751-4404), 32,680 ± 470 BP (Pta-2116), 31,870 ± 1,000 BP (Pta-2270), and 31,070 ± 750 BP (Pta-2361). Using the recently established calibration methods (Van Andel 1998), the plate is 32,500-38,000 calendar years old, including the confidence limits. There is some evidence that it belongs to the Aurignacian I/II period, 33,000-30,000 BC (Djindjian et al. 1999:165-166, 172).

The anthropoid was recognized as a 'lord of the animals' and a 'divine being' (Drössler 1991: 156) or a 'sky-god' (Hahn 1982: 11) that is involved in a ritual (Marshack 1987: 143). It was supposed that the anthropoid on side A of Fig. 1 is connected with the notations on side B and the edges, which indicates a sort of calendar, perhaps a lunar one, related to the solar year (Hahn 1982: 11; Bosinski

1982: 77-78, Boșinski 1990: 75; Müller-Beck 1985: 220-222; Rücklin 1995; Müller-Beck et al. 2001: 65-68). Since 1989, I have worked on the plate and propose here the hypothesis that the anthropoid represents Orion at the vernal equinox ca 31,000 BC (Rappenglück 1995) together with a combined lunar and pregnancy calendar, which is related to the heliacal rising and setting of Orion (Betelgeuse).

At first sight the anthropoid fits very well with Orion (as it looks today, i.e. the astronomical standard equinox of J2000.0), showing the sand-glass figure of the constellation (Krupp 1983: 8; Szyjewski 1999: 257): The straddle-legged posture of the anthropoid, with the right foot a little more highly raised in comparison with the left one, calls to mind the lower part of Orion. The narrow wasp's waist in particular designates the position of Orion's belt. The perpendicularly upraised arms, with the head between the hands, denote the upper part of Orion. Due to the great age of the small plate the proper motions of the stars were computed (using Starry Night Pro 3.12c) in steps of 1000 years back to 40,000 BC, which is the oldest available date of an Aurignacian layer in the cave (thermoluminescence OxA-4595, Richter et al 2000). The overall appearance of Orion to a stargazer 40,000-30,000 BC was almost the same as today (Fig. 2), except for the position of the star φ^2 Ori.[1] This star was then situated about 3° north of its present-day position (position angle 162.5°).

In the years 5,600-15,300 BC and 31,500-41,500 BC, Orion did not appear completely above the natural horizon at the Geißenklösterle cave. The small plate, however, shows a complete anthropoid. If it really represents Orion, then it is to be dated later than 5,600 BC or between 15,300 and 31,500 BC. With respect to the given ^{14}C accelerator-mass-spectrometry and thermoluminescence dating, the younger dates up to 30,000 BC can be excluded. Thus the small plate would belong to the epoch 30,000-31,500 BC.

Fig. 2. The 'Anthropoid in the Sky' culminating above the Geißenklösterle at the autumnal equinox, ca 30,000 BC. The precession of the equinoxes and the proper motion of the stars are taken into account (up to magnitude 6.0). Made using Starry Night Pro 3.12c by Michael A. Rappenglück. Image courtesy Starry Night Pro/SPACE.com.

[1] The notation of the stars in this study uses the common description with small Greek letters suggested by the German astronomer Johann Bayer (1572-1625); otherwise, if there doesn't exist one from Bayer, the number according to the British astronomer John Flamsteed (1646-1719).

The anthropoid shows an appendage between its legs, which looks like a stretched trapezoid with the wide side below. Hahn (1986: 119) sees here "a longish loincloth, an animal's tail or a penis". Such a loincloth can be recognized in a group of stars between σ Ori and $\theta^{1,2}$ Ori, ι Ori, including the Orion Nebula M 42 (Fig. 2). But Hahn's interpretation of the object as a penis is much more interesting. In the area of Orion many ancient people recognized a giant anthropoid, sometimes equipped with attributes of an animal and often seen as a hunter (Szyjewski 1999: 253-256). They related the constellation to the cosmic cycles of birth, life, death, and rebirth. Thus phallic, sexual, aquatic, and soteriological motives were connected with Orion (Szyjewski 1999; Krupp 1983: 22, 101; Sellers 1992). The Chukchee in Siberia, for example, saw Orion as Rulte'nnin, a male hunter with a crooked back who chases celestial animals (Werner 1952: 141). The star λ Ori marks his head; α Ori and γ Ori designate his left and right shoulders; κ Ori and β Ori denote his left and right legs. The belt stars (δ, ε, ζ Ori) shape the crooked back. Two bright stars, in particular ι Ori below the belt, denote his penis (Werner 1952: 141). The position of the constellation directly near the Milky Way, which people often recognized as a stream of milky semen, and the penis-like appendage—Orion's sword with the Orion Nebula (M 42)—supported the ideas of the ancients that Orion may be responsible for the celestial insemination of the earth and for cosmic fertility (Szyjewski 1999: 260). If, moreover, in the course of millennia, the points of the equinoxes or solstices had been situated near or above Orion, the star phases of selected stars of the constellation would have been used by people for timekeeping (see e.g. Lévi-Strauss 1976: 281-309; Griffin-Pierce 1998: 64; Sellers 1992).

Around 32,000 BC, the vernal equinox was situated above the head-star of Orion, φ^2 Ori (magnitude 4.06), only 0.5° from 121 Tau (magnitude 5.34). The ancients did not observe this point, but instead they watched the star phases of bright stars nearby. φ^2 Ori set heliacally nearly 22 days before the vernal equinox and rose close to the same number of days before the summer solstice. It remained invisible for about 95 days (Starry Night Pro 3.12c). This star together with λ and φ^1 Ori formed the ancient third Indian moon lodge *mṛga-śiras*—the head of a beast (a gazelle) (Scherer 1953: 155). This group was also called *agra-hayani*—the beginning of the year. In the *Suryaprajñapti*, a sacred book of the Jainas, it was named *samthana*—life, stand still, end, death. According to Vedic myth the asterism λ, φ^1, φ^2 Ori is the head of Prajapati (Varuna), signifying the year. This third, and also the fourth Indian moon lodge *Ardra* (Betelgeuse), marked the vernal equinox ca 4,500 BC (Allen 1963: 311, 319). There was a quite similar situation one cycle of precession before that date, at 30,000 BC. Betelgeuse (α Ori, magnitude 0.43v) was better situated for announcing springtime at ca 30,000 BC. It set heliacally about 14 days before the vernal equinox and rose heliacally nearly 19 days before the summer solstice (altitude 3°, Starry Night Pro 3.12c). Orion, then, could have been used as a celestial time signal announcing spring and summer. Betelgeuse remained invisible for 86 ± some days, depending on the natural horizon and atmospheric conditions. This reminds one of the sum of all notches of the Geißenklösterle plate, which is 87 ± 1.

A pregnancy lasts on average 280 days (40 weeks), counting the first day of the last normal menstrual period as day one (Brockhaus 1989: 193), i.e. about 266 days (38 weeks) after conception. Now nine synodic months are nearly 266 days (280-14 days) and 10 sidereal months are about 273 days (280-7 days), always counting with integers. A predictor to estimate the date of delivery, still used today by many doctors, is the so-called Naegele rule (Brockhaus 1989: 193): Starting with the first day of the last menstrual period, three calendar months (about 92 days) are subtracted and one year and seven days added. It is based on the concept that human gestation is 10 menstrual cycles of 28 days (nine sidereal months plus seven days) and works with 95% probability (confidence interval ± three weeks). The quantity of three calendar months is quite close to the above-mentioned one of 86 ± some days (and also to the notches on the plate, 87 ± 1), which is the time between the heliacal setting and rising of Orion (Betelgeuse) at the site of the Geißenklösterle cave and in the Aurignacian epoch. Thus the Palaeolithic timekeepers may have fixed the time of conception with the heliacal setting and predicted the birth after a rule similar to that of Naegele's. Perhaps at that time in the year a ritual was performed in which cosmic fertility was worshipped and sexual intercourse was preferred. Such traditions are well known from different ancient peoples all over the world (Szyjewski 1999: 258). Pregnancy calendars had been used as early as the Gravettian period, 29,000-22,000 BP (Duhard 1988: 23-39), and the 86 days, relating the solar year and the duration of pregnancy, played an important role in calendars of the Magdalenian era, 17,000-12,000 BP (Frolov 1974: 48-52, 63-64, 153-156).

Fig. 3. Three similar anthropoids from the area of the Danube Valley (Germany/Austria). Illustration: Michael A. Rappenglück.

There exist two quite similar artefacts, approximately of the same age, found at the Hohlenstein-Stadel, Lone valley, Germany, and at the Galgenberg, near Stratzing/Krems, Austria (Fig. 3). They show similar postures, but each of them adds a peculiar feature that is often a characteristic of Orion: One lifts an arm, or perhaps a club; the other has the head of a lion. Did some Aurignacian tribes in a small area along the Danube River have a common conception of an anthropoid, perhaps a feline, including the motive of a celestial humanoid and lion-like solar giant, who as a 'Lord of the Animals' hunted the stars? We do not know. But the find from the Geißenklösterle cave strengthens the hypothesis that around 30,000 BC some Aurignacian tribes had an idea of an anthropoid in the sky, which most likely was Orion.

*Bahnhofstrasse 1, 82205 Gilching-Geisenbrunn, Germany, mr@infis.org.

References

Allen, R. H. 1963. *Star Names. Their Lore and Meaning.* New York.

Bosinski, G. 1982. *Die Kunst der Eiszeit in Deutschland und in der Schweiz* (Römisch-Germanisches Zentralmuseum, Forschungsinstitut für Vor- und Frühgeschichte. Kataloge Vor- und Frühgeschichtlicher Altertümer 20). Bonn.

Bosinski, G. 1990. *Homo sapiens: L'histoire des chasseurs du Paléolithique supérieur en Europe (40000-10000 avant J.-C.).* Paris.

Bosinski, G. 1994. Menschendarstellungen der Altsteinzeit. In *Der Löwenmensch: Tier und Mensch in der Kunst der Eiszeit* (Begleitpublikation zu der Ausstellung im Ulmer Museum 11. September – 13. November 1994), 77-100. Sigmaringen.

Brockhaus Enzyklopädie (1989). S. v. *Geburt*, 192-194. Mannheim.

Djindjian, F., Koslowski, J. K., and Otte, M. 1999. *Le Paléolithique supérieur en Europe.* Paris.

Drössler, R. 1991. *Menschwerdung: Funde und Rätsel.* Leipzig, Jena, Berlin.

Duhard, J.-P. 1988. Le calendrier obstétrical de la femme à la corne de Laussel. *Bulletin de la Société Historique et Archéologique du Périgord* **65**, 23-39.

Frolov, B. A. 1974. *Zisla v grafike paleolita.* Novosibirsk.
Griffin-Pierce, T. 1998. *Earth is my mother, Sky is my father: Space, time, and astronomy in Navajo sandpainting.* Albuquerque.
Hahn, J. 1982. Eine menschliche Halbreliefdarstellung aus der Geissenklösterle Höhle bei Blaubeuren. In *Fundberichte aus Baden-Württemberg* 7, 1-12. Stuttgart.
Hahn, J. 1986. *Kraft und Aggression. Die Botschaft der Eiszeitkunst im Aurignacien Süddeutschlands? (Archaeologica Venatoria* 7). Tübingen.
Krupp, E. C. 1983. *Echoes of the ancient skies: The astronomy of lost civilizations.* New York, Oxford.
Lévi-Strauss, C. 1976. *Mythologica I: das Rohe und das Gekochte.* Frankfurt am Main.
Marshack, A. 1987. L'Évolution et la transformation du décor du début de l'aurignacien au magdalénien final. In *L'art des objets au paléolithique 2: Les voies de la recherche* (Actes de Colloque international Foix - Le Mas-d'Azil, 16-21 novembre 1987), 139-158. Paris.
Müller-Beck, H. 1985. Überlegungen zur Interpretation früher bildlicher Darstellungen. In *Jagen und Sammeln. Festschrift für Hans-Georg Bandi zum 65. Geburtstag (3. September 1985)*, ed. R. Fellmann, G. Germann, and K. Zimmermann, 217-224. Bern.
Müller-Beck, H., Conard, N. J., and Schürle, W. (ed.). 2001. *Eiszeitkunst im Süddeutsch-Schweizerischen Jura: Anfänge der Kunst.* Stuttgart.
Neugebauer-Maresch, C. 1990. Zum Neufund einer 30000 Jahre alten weiblichen Statuette bei Krems, Niederösterreich, *Antike Welt* **1**, 3-13.
Rappenglück, M. A. 1995. Unterwegs mit dem Sternbild im Handgepäck: Ein Fund aus der Höhle von "Geißenklösterle" bei Blaubeuren und Himmelsbeobachtungen vor 32000 Jahren. *Kultur Notizen* **15**, 5-20.
Richter, D., Waiblinger, J., Rink, W. J., and Wagner, G. A. 2000. Thermoluminescence, electron spin resonance and ^{14}C-dating of the Late Middle and Early Upper Palaeolithic site Geißenklösterle cave in Southern Germany. *Journal of Archaeological Science* **27**, 71-89.
Rücklin, G. 1995. Der Adorant – Ein Kalender aus dem Aurignacien? *Mitteilungsblatt der Gesellschaft für Urgeschichte* **3**, 8-10.
Scherer, A. 1953. *Gestirnnamen bei den indogermanischen Völkern.* Heidelberg.
Sellers, J. B. 1992. *The Death of Gods in Ancient Egypt: An Essay on Egyptian Religion and the Frame of Time.* London.
Starry Night Pro 3.12c. (SPACE.com software). 2000. Toronto.
Szyjewski, A. 1999. The Soteriological Context of Orion and Sirius Mythology in Tribal Traditions. In *Actes de la Vème Conférence Annuelle de la SEAC, Gdańsk 1997* (Światowit supplement series H: Anthropology 2), ed. A. LeBeuf and M. S. Ziółkowski, 251-261. Warsaw and Gdansk.
Van Andel, T. H. 1998. Middle and Upper Palaeolithic Environments and the Calibration of ^{14}C Dates beyond 10,000 BP, *Antiquity* **72**, 26-33.
Werner, H. 1952. Klassische Sternbilder am Himmel der Tschuktschen. *Zeitschrift für Ethnologie* **77**, 139-141.
Wingert, H. and Stephan, T. 1985. Das älteste Menschenbild ein Kalender? *Westermanns Monatshefte* **11**, 108-112.

The Ramesside star clocks and the ancient Egyptian constellations

Juan Antonio Belmonte*

Abstract

The ancient Egyptians had a complete set of constellations covering the whole sky, as seen from Egyptian latitudes. There were two dominants groups, one in the southern sky and another in northern declinations, both being separated probably by the Milky Way (Davis 1985) or by the Ecliptic (Krauss 1997). The northern sky, as beautifully represented in some ceilings of the tombs of the New Kingdom, was full of constellations, e.g. the Lion, the Crocodile, the Foreleg (sometimes a whole Bull), and the Hippopotamus. The southern group was essentially formed by a belt of 'constellations' known as the decans, individual stars or asterisms whose heliacal rising was used for timekeeping from the end of the Old Kingdom (ca 2200 BC), with Sah (parts of Orion) and Sepedet (Sirius) as the dominant ones. We know this group from the diagonal star clocks of the coffin lids of the First Intermediate Period and also from later tombs and temple ceilings, including the famous Zodiac of Denderah.

In the New Kingdom (ca 1500 BC), the decans were no longer useful for timekeeping, and a new system was developed using the meridian transit of certain stars, belonging in some cases to huge constellations such as the Female Hippopotamus or the Giant and a few asterisms. These star clocks have been found in the tombs of the last Ramesside pharaohs—hence the name—(ca 1100 BC) in the Valley of the Kings, where they were painted for the benefit of the deceased kings.

An analysis of the astronomical data presented in these clocks has allowed us to prepare a preliminary list of correlations between the Egyptian stars present in them and the actual stars in the sky. Some results are very coherent, such as the identification of the Lion with our constellation Leo or the identity between the Hippopotamus constellation of the northern sky representations in the tomb ceilings and the Female Hippopotamus of the Ramesside star clocks, an identification that had been questioned by several scholars in the past. Nevertheless, it is worth mentioning that the clocks have several internal inconsistencies and errors that are extremely difficult to explain, especially from the astronomical point of view and that might lead to alternative proposals. In any case these data have been used in combination with previous identifications of the decanal belt of stars to produce a hypothetical map of the ancient Egyptian firmament.

Introduction

> Just as in the case of the decans, the crudeness of the underlying procedures is so great that only under severely restrictive assumptions could numerical conclusions be abstracted from the given lists. If we add the fact of obvious errors and carelessness in details in the execution of the texts as we have them, one would do best to avoid all hypothetical structures designed to identify Egyptian constellations from the analysis of the Ramesside star clocks. (Neugebauer and Parker 1964: x).

This, in my opinion extremely unfortunate, paragraph was written in the early 1960's by two of the most reputable scholars in the field of ancient Egyptian astronomy. Despite their tremendous merits in the field (their *Egyptian Astronomical Texts*, published in three volumes in the 1960's, is a real masterpiece in most senses) statements like this have severely handicapped any advance in ancient Egyptian sky lore, especially when focused on an actual astronomical point of view. An advance would have been highly desirable especially because, with reference again to the Ramesside star clocks, such a "device seems to have appeared so profound and effective that it became destined to indicate the hours of night for all eternity on the ceilings of the Ramesside tombs" (Neugebauer and Parker 1964: 74).

However it is worth mentioning that some effort has been devoted to other areas of ancient Egyptian astronomy. General reviews can be found in Gallo (1998) and the excellent compilation by Clagett (1995). Discussions on the calendar, after the classic work of Parker (1950), can be found in Roy (1982), Wells (1994), and most recently in the interesting work of von Bomhard (1999). Besides, since the pioneering, and very controversial, book of Lockyer (1973 [1894]), Hawkins (1963; 1975),

Krupp (1979: 208-219; 1984; 1991), Haack (1984), Leitz (1991), Spence (2000), and Belmonte (1999, 2000, 2001a) have discussed archaeoastronomy and the problem of the astronomical orientation of monuments.

Regarding the skies themselves, apart from the early proposals of Petrie (1940) and Chatley (1940), where we already have the identifications of Meskhetiu (*msḫtiw*) with the Big Dipper, Sah (*s3ḥ*) with Orion and Sepedet (*spdt*) with Sirius, in the last two decades few attempts have been made to identify ancient Egyptian constellations, although we have interesting approaches to the problem using the mythological information contained in different sources, especially the Pyramid Texts of the fifth and sixth dynasties (Faulkner 1969). It is worth mentioning the attempts to identify the decans in the works of Böker (1984), Davis (1985), Sellers (1992), Locher (1981; 1983; 1985), Leitz (1995, specifically on the Ramesside star clocks), Krauss (1997), and my own (Belmonte 2001b).

Discussion

The Ramesside star clocks are found as tables on the ceilings of the tombs of Ramses VI, Ramses VII, and Ramses IX in the Valley of the Kings. They were not properly understood until the pioneering work of Le Page Renouf (1874), who first interpreted them as astronomical rather than astrological devices, as had been supposed by earlier (and later) scholars. He took the coordinate net seriously, gave a proper explanation of the procedure (although he took the tables as a calendar of astronomical observations and not as a star clock), and correctly established the date for which it was originally designed as about 1450 BC; Neugebauer and Parker (1964: 6) give a date of 1470 BC.

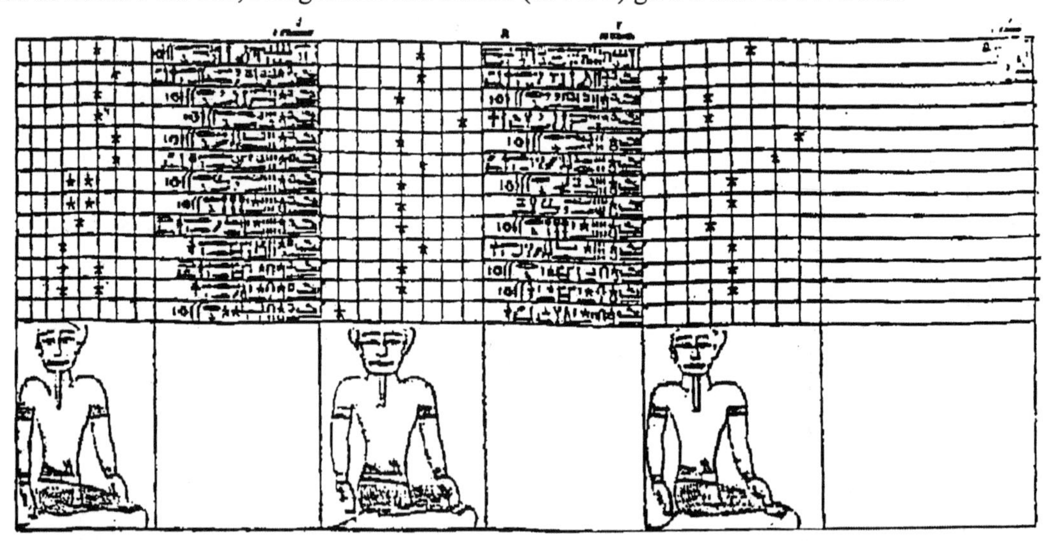

Fig. 1. Representation of the star clock in the corridor of the tomb of Ramses VI in the Valley of the Kings. Notice the hieroglyphic text with the hours of the night and the corresponding star, and the diagram with its relative position to the upper part of the hour-priest's body.

The tables in the tombs give the 12 nightly hours in a fortnightly scheme by the observation of certain stars and asterisms passing through a reference frame formed by the different parts of the body of a human image (Fig. 1). The positions are the following:

1. ḥr kꜥḥ i3by	On the left shoulder
2. ḥr msḏr i3by	On the left ear
3. ḥr irt i3byt	On the left eye
4. r ꜥk3 ib	In the middle
5. ḥr irt wnmyt	On the right eye
6. ḥr msḏr wnmy	On the right ear
7. ḥr kꜥḥ wnmy	On the right shoulder

All of the tables are summarised in Table 1 where for every fortnight in the Egyptian civil year different stars mark the beginning of the night and the 12 consecutive hours with positions referring to the previous list.

As can be seen in Table 2, the number of stars and asterisms used in the Ramesside clocks adds up to 47, although some appear only once while others are used often. Le Page Renouf also proposed several interpretations of the constellations in the clocks, e.g. Hadar as the 'goose head', the Pleiades as *ḫ3w*, Coma Berenices as the 'many stars', Aldebaran as the 'Star of Sar', and Arcturus and Antares as belonging to *mnit* (several of these identifications are confirmed by our proposal). However he would have liked to identify *sb3wy* (the 'Two Stars') with Castor and Pollux, but was unable to do so because he identified the 'Star of Sepedet' with Sirius and used it as the reference "fixed point to start from", exactly as proposed in Neugebauer and Parker several decades later (1964: 4).

Several other problems were identified by Neugebauer and Parker (1964) and by Clagett (1995). For them it was extremely difficult to establish the identity of the Mooring Post, the Hippopotamus, or the Lion in the Ramesside clocks, and the same constellations represented in the northern sky of several tomb and temple ceilings, particularly since, in the words of Clagett (1995: 63), "the stars of the Ramesside tables lay roughly parallel to but outside and south of the decanal belt". All these preconceived ideas made any identification extremely unreliable. For example, this obliges us to assume that 'on the left shoulder' means west of the meridian, which might not be necessarily correct. I agree, however, that 'in the middle' possibly marks the meridian crossing. Leitz (1995) assumed most of these preconceived ideas and proposed a complete identification of most of the stars in the Ramesside star clocks with faint and inconspicuous stars near the southern horizon, e.g. π Cet with *sb3 n ḫ3w*. For this work, consequently, I decided to ignore all of them. It is worth mentioning that, due to the lack of space, we will just present a rough description of our methodology. A complete description will be published elsewhere in the near future.

The first thing to do was to try to determine to which hour angle the difference between two positions in the 'human image' reference frame would correspond. By comparing those cases when a star or asterism jumps two hours in the tables instead of the standard one, it has been determined that one hour is approximately 2.3 ± 1.4 divisions. Consequently, we will consider for our purposes that one division is roughly one third of an hour, with a similar uncertainty of ~ ± 5° (20 minutes). Secondly, a quick glance at the tables clearly shows that the length of the hours is different for different times of the year and that it was extremely difficult to obtain internal congruency for all fortnights. Hence the starting point of the identification was to consider only the central hours of the night in those periods (close to the solstices and equinoxes) where it was possible to postulate a certain value for the length of the hours. We have taken 15° for the summer solstice, 12½° for the equinox, and 10° for the winter solstice hours respectively. The areas used in the identification are printed in boldface in Table 1.

Finally, a starting point is needed. Since we are not sure of the identification of the 'Star of Sepedet' with Sirius itself and since it is never used as an hour marker in our periods of reference, we have taken another, in our opinion clearer, identification, which is that *sb3 n ḫ3w* should be one of the stars of the asterism *ḫ3w*, the Pleiades cluster (Belmonte 2001b). However not all the stars are used as hour markers in the period of the solstices and the equinoxes, and we need also to connect one period with another. For these purposes certain diagonals were considered. Table 2 gives the result of the procedure where, for each star or asterism, a certain value of the right ascension has been derived. These values have been used to propose a series of possible identifications that have offered some interesting clues, such as the possibility that the Female Hippopotamus in the Ramesside clocks and the one in the north sky are the same, or the identification of the constellation of the Lion with our Leo and also with the lion in the ceiling representations. Interesting also is the identification of the 'Two Stars' with Castor and Pollux, as suspected by Le Page Renouf, or the 'Star of Water' with the Praesepe cluster, which has a misty aspect in the sky.

By the way, the 'Star of Spdt' lies between 21° and 37° (average of 28°) after the 'Star of Sah' and it is very probably Sirius, although this interpretation raises many problems. Besides, it should be almost at the same right ascension as its Follower since they are found together only once in the tables (II prt 16, see Table 1, Hour column, line 13) and while one is 'in the Middle' the other is 'on the left shoulder' (so we have around one hour plus and one hour minus, which returns us almost to the same position). This makes the identification of both stars very problematic.

Hour	0	I	II	III	IV	V	VI	VII	VIII	IX	X	XI	XII
I 3ḫt 1	šwty nt nḫt$_5$	tp n ḥḏ nt nḫt$_2$	nḥbt=f$_4$	bgs=f$_4$	sḏḥ=f$_7$	pt=f$_5$	ꜥryt$_4$	tp n 3pd$_4$	kft=f$_3$	sb3 n ḫ3w$_4$	sb3 n sꜥr$_4$	tpyꜥ s3ḥ$_4$	sb3 n s3ḥ$_1$
I 3ḫt 16	tp nḫt$_5$	nḥbt=f$_5$	bgs=f$_4$	sḏḥ=f$_7$	pt=f$_4$	ꜥryt$_5$	tp n 3pd$_4$	kft=f$_4$	sb3 n ḫ3w$_4$	sb3 n sꜥr$_5$	tpyꜥ s3ḥ$_4$	sb3 n s3ḥ$_4$	sb3 n spdt$_1$
II 3ḫt 1 Equinox	nḥbt nḫt$_4$	bgs=f$_5$	sḏḥ=f$_4$	pt=f$_4$	ꜥryt$_5$	tp n 3pd$_5$	kft=f$_4$	sb3 n ḫ3w$_4$	sb3 n sꜥr$_3$	tpyꜥ s3ḥ$_3$	sb3 n s3ḥ$_3$	sb3 n spdt$_5$	tpyꜥ sb3wy$_4$
II 3ḫt 16	bgs n nḫt$_4$	pd nḫt$_4$	pt=f$_4$	ꜥryt$_5$	bꜥnt n 3pd$_3$	kft=f$_4$	sb3 n ḫ3w$_5$	sb3 n sꜥr$_5$	tpyꜥ s3ḥ$_4$	sb3 n s3ḥ$_7$	iy ḥr-s3 sb3 n spdt$_1$	tpyꜥ sb3wy$_1$	sb3w nw mw$_4$
III 3ḫt 1	pd nḫt$_4$	pt=f$_4$	ꜥryt$_5$	tp n 3pd$_5$	kft=f$_4$	sb3 n ḫ3w$_4$	sb3 n sꜥr$_4$	tpyꜥ s3ḥ$_4$	sb3 n s3ḥ$_3$	iy ḥr-s3 sb3 n spdt$_5$	tpyꜥ sb3wy$_4$	sb3w nw mw$_4$	tp m3i$_4$
III 3ḫt 16	pt$_4$	ꜥryt$_5$	tp n 3pd$_4$	kft=f$_4$	sb3 n ḫ3w$_4$	sb3 n sꜥr$_3$	tpyꜥ s3ḥ$_4$	sb3 n s3ḥ$_5$	iy ḥr-s3 sb3 n spdt$_5$	tpyꜥ sb3wy$_4$	sb3w nw mw$_4$	tp m3i$_4$	sd=f$_4$
IV 3ḫt 1	sb3 n s3 pt$_3$	tp n 3pd$_3$	kft=f$_5$	sb3 n ḫ3w$_7$	sb3 n sꜥr$_7$	sb3 n s3ḥ$_3$	iy ḥr-s3 n spdt$_4$	tpyꜥ sb3wy$_1$	sb3wy$_5$	sb3w nw mw$_6$	tp m3i$_7$	sd=f$_7$	sb3w ꜥš3w$_5$
IV 3ḫt 16	tp n 3pd$_4$	kft=f$_4$	sb3 n ḫ3w$_6$	sb3 n sꜥr$_6$	sb3 n s3ḥ$_4$	sb3 n spdt$_7$	sb3wy$_1$	sb3w nw mw$_3$	tp m3i$_3$	sd=f$_3$	sb3w ꜥš3w$_4$	ṯ3 nfr$_4$	šmsw n ḥ3t n mnit$_5$
I prt 1 Winter Solstice	kft=f$_4$	sb3 n sꜥr$_3$	tpyꜥ sb3 n s3ḥ$_4$	sb3 n s3ḥ$_5$	iy ḥr-s3 n spdt$_3$	sb3wy$_2$	sb3w nw mw$_2$	tp m3i$_3$	sd=f$_4$	sb3w ꜥš3w$_5$	šmsw n ḥ3t n mnit$_1$	mnit$_3$	šmsw mnit$_3$
I prt 16	tpyꜥ sb3 n s3ḥ$_1$	sb3 n s3ḥ$_1$	sb3 n spdt$_4$	tpyꜥ sb3wy$_4$	sb3wy$_2$	sb3w nw mw$_3$	tp m3i$_4$	sd=f$_5$	sb3w ꜥš3w$_4$	šmsw n ḥ3t n mnit$_1$	mnit$_3$	šmsw mnit$_1$	pd n rrt$_1$
II prt 1	sb3 n s3ḥ$_1$	sb3 n spdt$_6$	(tpyꜥ) sb3wy$_6$	sb3w nw mw$_3$	tp m3i$_2$	sd=f$_2$	sb3w ꜥš3w$_2$	ṯ3 nfr$_4$	šmsw (n) ḥ3t mnit$_4$	mnit$_4$	rd n rrt$_3$	pd=s$_3$	mnty=s$_4$
II prt 16	sb3 n spdt$_4$	iy ḥr-s3 n spdt$_7$	(tpyꜥ) sb3wy$_6$	tp m3i$_1$	sd=f$_3$	sb3w ꜥš3w$_4$	ṯ3 nfr$_4$	šmsw (n) ḥ3t mnit$_4$	mnit$_2$	šmsw mnit$_4$	(pd n) rrt$_2$	ḥry-ib n mnty =s$_4$	ḫpd=s$_4$
III prt 1	(tpyꜥ) sb3wy$_7$	sb3w nw mw$_3$	(tp n m3i)$_4$	s(d=f)$_5$	sb3w ꜥš3w$_4$	(ṯ3 nfr)$_4$	(šmsw n ḥ3t mnit)$_3$	šmsw mnit$_3$	rd n rrt$_3$	pd=s$_4$	ḥry-ib n mnty =s$_4$	ḫpd=s$_4$	mndt= s$_4$
III prt 16	sb3w nw mw$_4$	m3i$_5$	sd=f$_4$	sb3w ꜥš3w$_4$	ṯ3 nfr$_4$	šmsw ḥ3t n mnit$_4$	mnit$_7$	rdwy n rrt$_5$	pd=s$_5$	ḥry-ib n mnty =s$_4$	ḫpd=s$_7$	mndt= s$_7$	šwty=s$_4$
IV prt I Equinox	tp m3i$_4$	sd=f$_3$	sb3w ꜥš3w$_3$	ṯ3 nfr$_5$	šmsw ḥ3t mnit$_5$	šmsw iy ḥrs3 mnit$_3$	rdwy n rrt$_3$	pd=s$_5$	ḥry-ib n mnty =s$_4$	ḫpd=s$_4$	mndt= s$_6$	ns=s$_7$	tpyꜥ šwty nt nḫt$_6$
IV prt 16	(sd n) m3i$_5$	sb3w ꜥš3w$_3$	ṯ3 nfr$_5$	šmsw ḥ3t mnit$_5$	šmsw iy mnit$_3$	rdwy n rrt$_3$	pd=s$_4$	b3ḥ=s$_4$	ḫpd=s$_7$	mndt= s$_6$	ns=s$_7$	tpyꜥ šwty nt nḫt$_6$	šwty nt nḫt$_4$
I šmw 1	tpyꜥ mnit$_6$	šmsw ḥ3t n mnit$_3$	mnit$_3$	šmsw mnit$_3$	rdwy n rrt$_4$	pd=s$_4$	b3ḥ=s$_6$	ḫpd=s$_7$	mndt= s$_6$	šwty=s$_6$	tpyꜥ šwty nt nḫt$_6$	šwty nt nḫt$_2$	tp ḥḏ nt nḫt$_2$
I šmw 16	šmsw n (ḥ3t n mnit)$_3$	mnit$_5$	šmsw mnit$_3$	rdwy n rrt$_4$	pd=s$_4$	b3ḥ=s$_7$	ḫpd=s$_7$	mndt= s$_6$	šwty=s$_6$	tpyꜥ šwty nt nḫt$_6$	šwty nt nḫt$_2$	tp ḥḏ nt=f$_4$	ḫ3b=f$_7$
II šmw 1	(šmsw) n mnit$_4$	rdwy n rrt$_4$	pd=s$_3$	b3ḥ=s$_4$	ḫpd=s$_3$	mndt= s$_3$	ns=s$_3$	šwty=s$_6$	tpyꜥ šwty nt nḫt$_6$	šwty nt nḫt$_4$	ḫ3b=f$_4$	mndt=f$_4$	bgs=(f)$_4$
II šmw 16	rdwy n rrt$_4$	pd=s$_4$	b3ḥ =s$_4$	ḫpd=s$_4$	mndt= s$_5$	ns=s$_5$	šwty=s$_6$	(šwty) nt nḫt$_2$	tp=f$_4$	ḫ3b=f$_4$	mndt=f$_7$	bgs=(f)$_4$	pd=f$_4$
III šmw 1 Summer Solstice	b3ḥ n rrt$_5$	ḫpd =s$_4$	mndt= s$_3$	ns=s$_4$	šwty=s$_4$	tpyꜥ šwty nt nḫt$_4$	šwty nt nḫt$_5$	nḥbt=f$_4$	mndt=f$_4$	(bgs)=f$_4$	pd=f$_3$	sbḳ=f$_4$	pt=f$_4$
III šmw 16	ḫpd n rrt$_4$	mndt= s$_4$	ns=s$_4$	šwty =s$_4$	tpyꜥ šwty nt nḫt$_4$	šwty nt nḫt$_5$	ḫ3b=f$_4$	mndt=f$_4$	bgs=f$_4$	pd=f$_4$	sbḳ=f$_3$	iy s3 pt=f$_3$	ꜥryt$_4$
IV šmw 1 (wrong)	mndt nt rrt$_6$	tpyꜥ šwty nt (nḫt)$_4$	ḥḏ nt nḫt$_4$	ḫ3b=f$_4$	bgs=f$_5$	pd=f$_4$	sbḳ=f$_4$	pt=f$_5$	ꜥryt$_7$	ḥtyt nd 3pd$_7$	kft=f$_7$	sb3 n ḫ3w$_7$	sb3 n s3ḥ$_4$
IV šmw 16	šwty nt rrt$_6$	šwty nt (nḫt)$_4$	..........$_4$	nḥ(bt=f)$_4$	mn(dt=f)$_3$	bgs=(f)$_2$	s(bḳ=f)$_7$	(pt)=f$_5$	ꜥryt$_4$	(tp n) 3pd$_6$	(kft)=f$_5$	sb3 n ḫ3w$_3$	sb3 n (sꜥr)$_4$

Table 1. Hours of the Night in the Ramesside Star Clocks. Areas used for the analysis are in boldface. Occasionally, diagonals are also used, e.g. *sb3w ꜥš3w* & *ṯ3 nfr* or *tpyꜥ s3ḥ* & *sb3 n s3ḥ*, to establish the angular distance between stars or asterisms.

Epoch	Stars & asterisms	No.	Const.	Translation	α (° / h)	Proposed identification (α in h)
	tpyʿ šwty nt nḫt	7	*nḫt?*	Predeccesor of the giant's 2 feathers	260 / 17.3	Altair (17.1)
I 3ḫt 1	*šwty nt nḫt*$_4$	9	*nḫt*	The 2 feathers of the giant	273 / 18.4	Sualocin -αDel- (18.1)
  I 3ḫt 16	*tp n ḥḏ nt nḫt* *ḥḏ(t) nt nḫt* *tp nḫt*$_5$	3 1 2	*nḫt* *nḫt* *nḫt*	Head of mace (or Crown) of giant Mace (or Crown) of the giant Head of the giant		Sadalsuud (18.5) Area of Aquarius? Area of Equuleus
II 3ḫt 1 E	*ḫ3b=f* *Nḥbt nḫt*$_4$	5 5	*nḫt* *nḫt*	Nape of his neck Neck of the giant	280 / 18.7 280 / 18.7	Enif -ε Peg- (18.8) Enif -ε Peg- (18.8)
	mndt=f	5	*nḫt*	His breast	290 / 19.3	θPeg (19.2) or ιPeg (19.5)
II 3ḫt 16	*bgs n nḫt*$_4$	9	*nḫt*	Hip of the giant	300 / 20.0	Markab -α Peg- (20.2)
 III 3ḫt 1	*sḏḥ=f* *pd nḫt*$_4$	3 6	*nḫt* *nḫt*	His shank Knee of the giant	304 / 20.2 309 /20.6	Scheat -β Peg- (20.3) Area of the Square of Pegasus
	sbḳ=f	3	*nḫt*	His foot	317 / 21.1	αAnd (21.3)
III 3ḫt 16	*pt*$_4$ *iy s3 pt=f*	9 1	*nḫt* *nḫt?*	Pedestal The one coming after his pedestal	317 / 21.1 327 / 21.8	αAnd & γ Peg (21.3) δAnd (21.8)
IV 3ḫt 1	*sb3 n s3 pt*$_3$ *ʿryt*	1 9	*nḫt?* *ʿryt*	★of the back of the pedestal Jaws (or Rising ★★★)	 333 / 22.2	Area of Andromeda Cassiopeia with Schedar (21.9)
IV 3ḫt 16	*bʿnt nt 3pd* *tp n 3pd*$_4$ *ḥtyt nt 3pd*	1 8 1	*3pd* *3pd* *3pd*	Peak of the bird Head of the bird Throat of the bird	337 / 22.5 346 / 23.1	αTri (22.8) βTri (23.1) Area of Algol
I prt 1 WS	*kft=f*$_4$	11	*3pd*	Its rump	354 / 23.6	αPer (23.9)
.	*sb3 n ḫ3w*	10	*ḫ3w*	Star of the flock or Miriad	9 / 0.6	Pleiades (0.6)
	sb3 n sʿr	10	*sb3 n sʿr*	Star of fire or arising	19 / 1.3	Capella (1.4) or Aldebaran (1.5)
I prt 16	*ṯpyʿ sb3 n s3ḥ*$_1$	8	*s3ḥ*	Predeccesor of Sah ★	29 / 1.9	Bow of Orion with πOri (1.9)
II prt 1	*sb3 n s3ḥ*$_1$	12	*s3ḥ*	★ of Sah	32 / 2.2	Rigel (2.5) or Orion's Belt (2.7)
II prt 16	*sb3 n spdt*$_4$ *iy ḥr-s3 sb3 n spdt*	6 6	*spdt* *spdt*	★ of Spdt (the Triangle) One coming "after" the ★ of Spdt	60 / 4.0 45 / 3.0	Sirius (4.2) ? * Betelgeuse (2.9) ?
III prt 1	*(tpyʿ) sb3wy*$_7$	9	*sb3wy*	(Predeccesor of) the pair of stars	50 / 3.3	Alhena -γGem- (3.3)
	sb3wy	4	*sb3wy*	The Pair of stars	55 / 3.7	Castor (3.8) & Pollux (4.1)
III prt 16	*sb3w nw mw*$_4$	10	*mw*	★★★ of water	75 / 5.0	Praesepe Cluster –M44- (5.2)
IV prt I E	*tp m3i*$_4$	11	*m3i*	Head of the lion	95 / 6.3	Regulus (6.8) & Head of Leo
IV prt 16	*sd n m3i*$_5$	11	*m3i*	Tail of the lion	115 / 7.7	Duhr -δLeo- (7.8) & Back of Leo
	sb3w ʿš3w	10	*m3i?*	Many stars	135 / 9.0	Coma B. (9.3 to 9.5) or γCor (9.3)
 I šmw 1	*ṯ3 nfr* *tpyʿ mnit*$_6$	7 1	*mnit?* *mnit*	Beautiful sprout Predeccesor of the Pole	150 / 10.0	Mizar (10.1) or Spica (10.4) Area of Ursa Maior & Bootes **
I šmw 16	*šmsw n (ḥ3t n mnit)*$_3$	11	*mnit*	Follower of the front of the pole	162 / 10.8	ηBoo (11.0) **
	mnit	7	*mnit*	Pole or Mooring post	165 / 11.0	Alkaid (11.0) **
***	*šmsw iy ḥr-s3 mnit*	2	*mnit*	Follower which comes after the pole	167 / 11.1	Alkaid (11.0) or Arcturus (11.5) **
II šmw 1	*(šmsw) n mnit*$_4$	7	*mnit*	(Follower of) the pole	171 / 11.4	Arcturus (11.5)
 II šmw 16	*rd n rrt* *rdwy n rrt*$_4$	2 7	*rrt* *rrt*	Foot of the hippopotamus Feet of the hippopotamus	 180 / 12.0	ρBoo (11.9) Izar -εBoo- (12.1) & ρBoo (11.9)
	pd n rrt	11	*rrt*	Knee of the hippopotamus	198 / 13.2	Gemma (13.5)
 III šmw 1 SS	*ḥry-ib mnty=s* *b3ḥ n rrt*$_5$	5 6	*rrt* *rrt*	Middle of her thighs Vulva of the hippopotamus	206 / 13.7 211 / 14.0	Gemma (13.5) & CrB Between Corona Bor & Herculis
III šmw 16	*ḥpd n rrt*$_4$	11	*rrt*	Buttocks of the hippopotamus	219 / 14.6	εHer (14.8) & ζHer (14.5)
IV šmw 1	*mndt nt rrt*$_6$	11	*rrt*	Breast of the hippopotamus	228 / 15.2	πHer (15.3)
	ns=s	6	*rrt*	Her tongue	240 / 16.0	Rastaban - βDra- (16.2)
IV šmw 16	*šwty nt rrt*$_6$	8	*rrt*	The 2 feathers of the hippopotamus	250 / 16.7	Lyra with Vega (16.7)

Table 2. Possible identification of stars in the Ramesside star clocks. The table presents the time of the year when the star marked the beginning of the night, the name of the star and asterism and its translation, the number of times it appears in the tables (a possible indication of its importance), the constellation it belongs to, the approximate right ascension—both in degrees and hours, and finally the proposed identification, allowing an error of ± 5° or 20 minutes in right ascension for 1500 BC. Those numbers resulting from the areas in boldface in Table 2 are underlined. Those stars which are interchangeable because they never appear one after the other in the tables are found in the same row without a separating line between them. The right ascension of Sirius (*) does not fit the general pattern (see text for further discussion). The constellation *mnit* (**) also has some uncertainties. *šmsw iy ḥr-s3mnit* (***) can substitute for either *mnit* or *šmsw n mnit*.

No.	✳	Constellation	Translation	Comments
1		*s3ḥ*	Sah (Creator)	South of Orion (Locher 1983)
2		*spdt*	Triangle	Sirius & its companions
3		*knmt*	Cow or ...	An alternative to the human figure of Sothis Canis Major
4		*iy ḥr-s3 sb3 n spdt*	Follower of the Triangle	Procyon (4) follows Sirius but Betelgeuse (4') fits RA
5		*tpyʿ sb3wy*	Predecessor of the Two Stars	Alhena, in Gemini
6		*sb3wy*	Pair of Stars	The Twins, Castor & Pollux
7		*sb3w nw mw*	Star of Water	The Praesepe Cluster (M44)
8		*ḏ3t*	The Ferryboat	Argo Navis. *ẖ3w* in Middle Kingdom.
9		*š3k*	Sek Crocodile	Our Hydra
10		*ṯm3t(y)*	The Two Wings	In the area of Argo Navis
11		*M3i*	The Lion	Our Leo
12		*šb3w ʿš3w*	Many Stars	Cumuli at Coma (12) or area of Corvus (12')
13		*wš3t bk3ty*	The Twins & the Two Pregnants	Southern Cross
13'		*ipḏs - sbšsn*	Its Own Count & the Sage's Star ?	βCentauri & αCentaruri, respectively
14		*šrkt*	Selkis Goddess	Our Virgo. Spica. It might be the star *ṯ3 nfr*
15		*ẖntt*	The Front	Perhaps the scorpion of some tomb ceilings
16		*ḳdty*	The Two Nets	One of them would be Corona Australis
17		*ẖnwy*	2 Khanuwy Fishes	αSgr & βSgr
18		*wi3*	The Boat	Capricornus. It might extend until Sagittarius
19		*srt*	The Sheep/Goat	In the area of Grus
20		*nẖt*	The Giant	From Aquila to the Pegasus Square. Including his Feathers, Mace (or Crown) and Pedestal
21		*3ẖwy*	The Two Spirits	Faint Stars in the Area of Sculptor
22		*b3wy*	The Two Souls	Stars in Cetus
23		*ḳd*	The Sheepfold	Head of Cetus
24		*ẖ3w*	Myriad or Flock	The Pleiades Cluster

25		*ʿrt*	Jaw	The Hyades Cluster
26		*sb3 n sʿr*	Star of fire or ascending star	Capella (most probable) or Aldebaran
27		*3pd*	The Bird	Triangulum and Perseus
28		*ʿryt*	(The Two) Jaw(s)	Cassiopeia
29		*ʿn(w)*	The God Horus like the Harpooner	There are two alternatives, 29 (most probable and inspired on paintings) & 29'
30		*msḫtiw*	The Thigh or The Percussor	In dotted-line, one of the Heaven's Adzes Mizar might be *ṯ3 nfr* instead of Spica
31		*mnit*	Mooring Post	The Star *mnit* might be Alkaid
32		*mnitwy*	The Two Poles (2 Mooring Posts)	The Celestial and the Ecliptic Poles? In dot-line, the second of the Heaven's Adzes
33		*rrt (3st ḏ3t mwt hb pt)*	The Hippopotamus	Big area near the Pole covering from Boötes to Lyra
34		*msḥ ?*	A 2nd Crocodile	The Crocodile on the back of *rrt*
35				A beheaded animal in Zodiac of Denderah (highly speculative). Area of Ophiuchus
			Standing man/god	Probably the air god Shu, but perhaps equal to number 20
36		*nwt*	The Goddess Nwt	She would be the Milky Way, according to Wells (1994)

Table 3. Ancient Egyptian stars and asterisms. For every constellation the table presents a number (see Fig. 2), its name in hieroglyphic and the transcription, its name in English, and the identification with modern constellations and stars. The constellation *msḫtiw*, when represented by a complete bull, might extend to the area of Camelopardalis or Leo Minor.

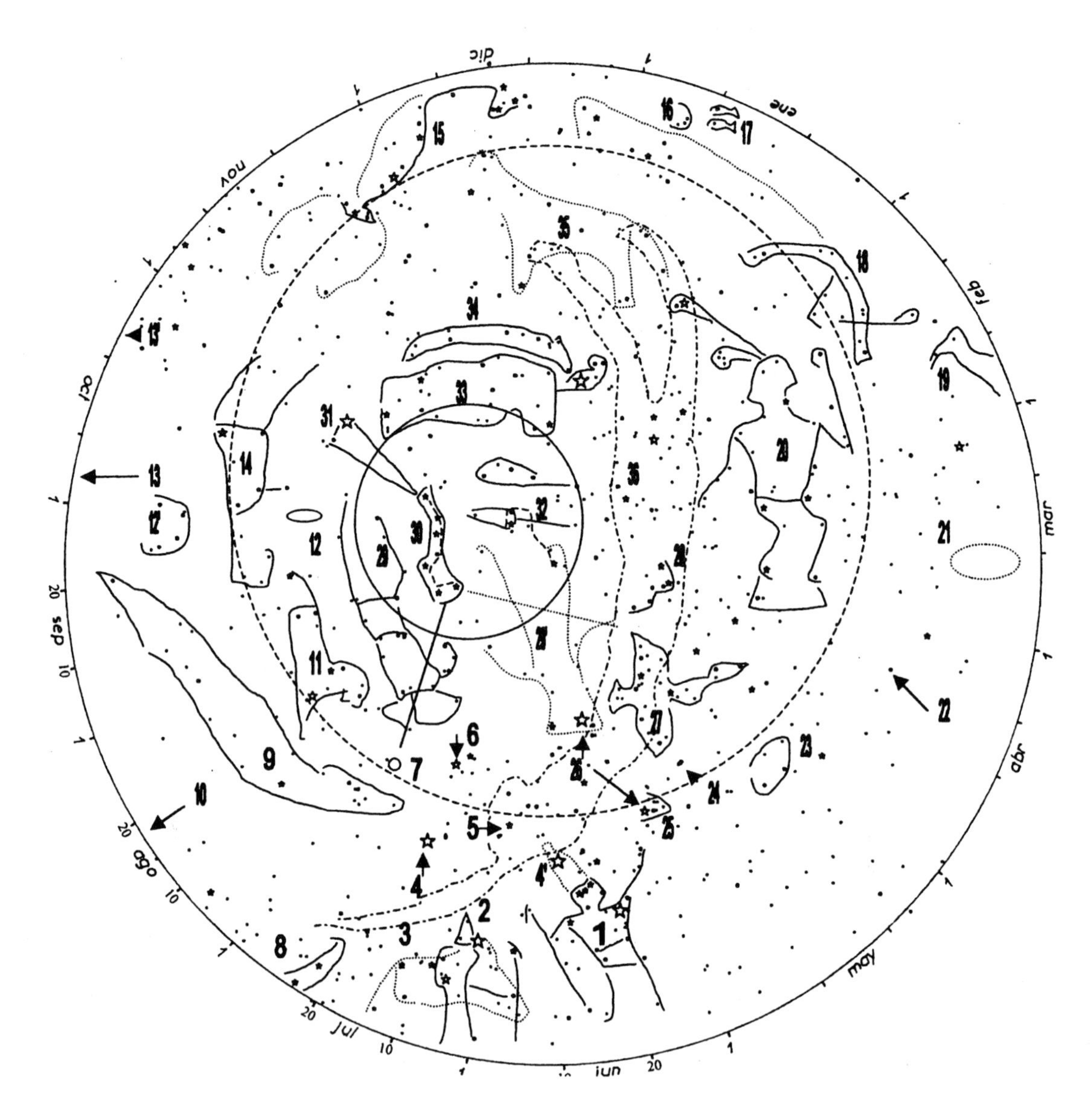

Fig. 2. The ancient Egyptian firmament. Stars and constellations are identified by numbers as established in Table 3. The Ecliptic and the circumpolar region are also plotted.

Conclusions

The data presented in this paper together with those derived for the decans in Belmonte (2001b) have been used to produce a 'chart' of the Egyptian firmament as presented in Table 3 and plotted in Fig. 2. I know that this proposal still has several problems and uncertainties that need to be explained and that other reasonable alternatives could easily be worked out. For example, we only agree with Leitz´s approach to the Ramesside star clocks in the presumable identification of *sb3 n s3ḥ* with Rigel (1995: 256). However, in so doing, I hope to have re-opened a way that was closed half a century ago and to motivate serious scholars to pursue the study of ancient Egyptian astronomy and sky lore.

*Instituto de Astrofísica de Canarias, Via Láctea S.N., La Laguna 38200 Tenerife, Spain, jba@ll.iac.es.

References

Belmonte, J. A. 1999. *Las Leyes del Cielo*. Madrid.
Belmonte, J. A. 2000. Astronomía y arquitectura: el papel de los astros en la cultura y el arte del antiguo Egipto. In *Arte y Sociedad del Antiguo Egipto*, ed. M. A. Sola and D. Sola, 109-35. Madrid.
Belmonte, J. A. 2001a. On the orientation of the Old Kingdom Egyptian pyramids. *Archaeoastronomy* (Supplement to the *Journal for the History of Astronomy*) **26**, S1-S20.
Belmonte, J. A. 2001b. The decans and the ancient Egyptian sky-lore. To appear in the proceedings of the INSAP III meeting, Palermo, 30 December-6 January 2001. *Memorie della Scieta Astronomica Italiana.* In press.
Böker, R. 1984. Über Namen und Identifizierung der ägyptischen Dekane. *Centaurus* **27**, 189-217.
von Bomhard, A. S. 1999. *The Egyptian calendar: A work for eternity*. London.
Chatley, H. 1940. Egyptian astronomy. *Journal of Egyptian Archaeology* **26**, 120-126.
Clagett, M. 1995. *Ancient Egyptian science*, vol. 2: *Calendars, clocks, and astronomy* (Memoirs of the American Philosophical Society 214). Philadelphia.
Davis, V. L. 1985. Identifying ancient Egyptian constellations. *Archaeoastronomy* (Supplement to the *Journal for the History of Astronomy*) **9**, S102-S104.
Faulkner, R. O. 1969. *The ancient Egyptian pyramid texts*. Oxford.
Gallo, C. 1998. *L'astronomia egizia*. Padova.
Haack, S. C. 1984. The Astronomical orientation of the Egyptian pyramids. *Archaeoastronomy* (Supplement to the *Journal for the History of Astronomy*) **7**, S119.
Hawkins, G. S. 1973. *Beyond Stonehenge*. London and New York.
Hawkins, G. S. 1975. Astroarchaeology: The unwritten evidence. In *Archaeoastronomy in pre-columbian America*, ed. A. Aveni, 131-162. Austin.
Krauss, R. 1997. *Astronomische konzepte und jenseitsvorstellungen in den pyramidentexten* (Ägyptologische Abhandlung 59). Wiesbaden.
Krupp, E. C. 1979. *In search of ancient astronomies*. London.
Krupp, E. C. 1984, Egyptian astronomy: A tale of temples, traditions, and tombs. In *Archaeoastronomy and the roots of science* (American Association for the Advancement of Science selected symposium 71), ed. E. C. Krupp, 289-320. Boulder.
Krupp, E. C. 1991. *Beyond the blue horizon*. Oxford.
Leitz, C. 1991. *Studien zur Ägyptischen Astronomie* (Ägyptologische Abhandlungen 49). Wiesbaden.
Leitz, C. 1995. *Altägyptische Sternuhren* (Orientalia Lovaniensia Analecta 62). Leuven.
Le Page Renouf, P. 1874. Calendar of astronomical observations found in royal tombs of the XXth dynasty. *Transactions of the Society of Biblical Archaeology* **3**, 400-421.
Locher, K. 1981. A conjecture concerning the early Egyptian constellation of the Sheep. *Archaeoastronomy* (Supplement to the *Journal for the History of Astronomy*) **3**, S63-S65.
Locher, K. 1993, New arguments for the celestial location of the decanal belt and the origins of the *s3h* hieroglyph. In Atti di *sesto congresso internazionale di egittologia*, vol. 2, 279-280. Turin.
Locher, K. 1985. Probable identification of the ancient Egyptian circumpolar constellations. *Archaeoastronomy* (Supplement to the *Journal for the History of Astronomy*) **9**, S152-S153.
Lockyer, J. N. 1973. *The Dawn of astronomy*. Cambridge MA (reprint of the 1894 edition).
Neugebauer, O. and Parker, R. A. 1964. *Egyptian astronomical texts*, vol. 2: *The Ramesside star clocks*. Providence.
Parker, R. A. 1950. *The calendars of ancient Egypt*. Chicago.
Petrie, W. M. F. 1940. *Wisdom of the Egyptians*. London.
Roy, A. E. 1982, The Astronomical basis of Egyptian chronology. *Society for Interdisciplinary Studies Review* **6**, 53-55.
Sellers, J. B. 1992. *The death of gods in ancient Egypt: An essay on Egyptian religion and the frame of time*. London.
Spence, K. 2000. Ancient Egyptian chronology and the astronomical orientation of pyramids. *Nature* **408**, 320-324.
Wells, R. A. (1994). Re and the calendars. In *Revolutions in time: Studies in ancient Egyptian calendrics*. (Varia Aegyptiaca Supplement 6), ed. A. J. Spalinger, 1-37. San Antonio.

The archaeoastronomy of the Palaikastro kouros from Crete

Robert Hannah* and Marina Moss**

Abstract

The Palaikastro kouros is an ivory and gold statuette of a young man that was discovered in Palaikastro in Crete in 1987-1990. It has been interpreted by the excavators as a representation of a deity similar to the Egyptian god Osiris and associated with the constellation Orion. They argue that this god was understood as rising from the dead at harvest-time at the end of the agricultural year, the time when Orion rose heliacally each year.

This paper analyses the astronomical arguments put forward in support of the excavators' hypothesis and finds them wanting on both astronomical and iconographical grounds. For the iconography the authors compare the kouros with both Cretan and Egyptian works of art. For the astronomy they consider the neighbouring calendrical and cosmographical systems of Egypt and Babylonia. In no case do they find it possible to agree with the current interpretation of the kouros.

The authors agree that the kouros probably represents a god and that his image reflects Egyptian influence. However comparison with a contemporary seal stone from Kydonia in Crete and with evidence from Egypt suggests that he is not connected with a constellation, but with the sun itself.

Discussion

Between 1987 and 1990, hundreds of fragments of a statuette (Fig. 1) were discovered at Palaikastro, in east Crete. The excavators have suggested that the statuette was housed in the small room beside Room 1 in Building 5 (MacGillivray et al. 1991: 129). In Room 2, which adjoins it, there was a cist (MacGillivray et al. 2000: 27). Broken pottery found in the room and within the cist dates the fire that swept the site to Late Minoan IB (ca 1600/1580-1500/1490 BC).

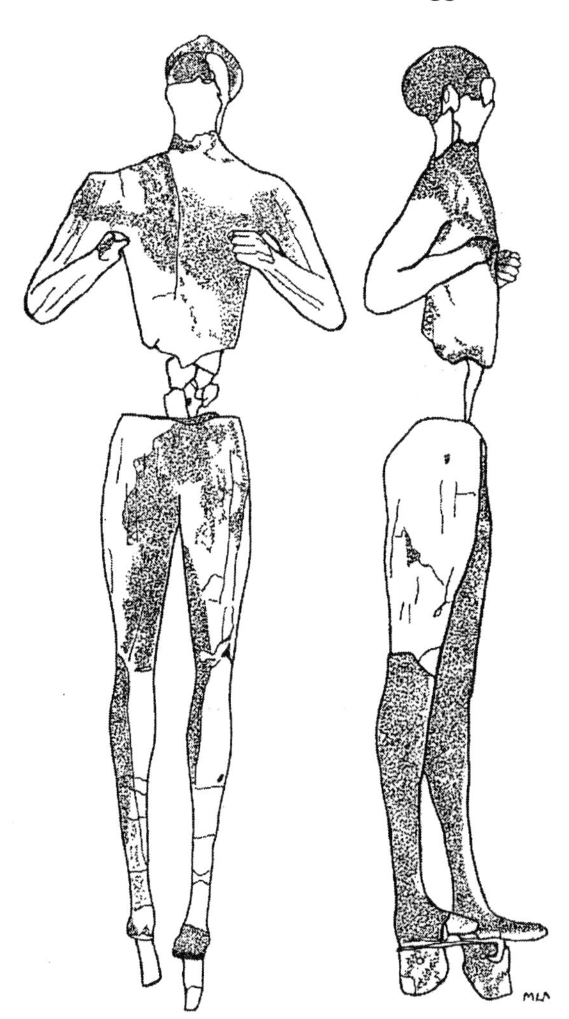

Fig. 1. The Palaikastro kouros (after MacGillivray et al. 2000: frontispiece, with permission).

The statuette is termed a kouros because of its association with the site of Palaikastro, where a fragmentary hymn to "the Greatest Kouros" was discovered in 1904 (Perlman 1995: 161-167). This version of the hymn is a second or third-century AD copy of an original of the late fourth or third century BC. The hymn was to be sung at the annual festival of Zeus Diktaios to whom a sanctuary had been built at Palaikastro probably in the eighth century BC, to the southeast of the Minoan site where the statuette was discovered.

The Palaikastro kouros has its left foot extended in front of the right. This is a feature not found on most male figurines from Minoan Crete, but it is a convention used in Egyptian art over thousands of years. The kouros holds his arms bent upwards to his chest, the fists clenched. In Egyptian art the hands placed in such a pose would have been given attributes to hold, but this appears not to have been the case with the kouros. The pose is not common in Cretan sculpture, but it is a feature of figurines from the neighbouring peak sanctuary sites of Petsophas and possibly Traostalos (Henriksson and Blomberg 1996: 113). These have usually been identified as votive offerings rather than images of a deity. But the rarity of the pose among Minoan figurines, and the association between Petsophas in particular (as the site of Mt Dikte) and the Greek god Zeus (MacGillivray et al. 2000: 145-148) suggest that the figurines from these sites may be representations of a deity. In addition, Rooms 1 and 2 in Building 5 could have been used for ritual (MacGillivray et al. 2000: 87-90). The cist in Room 2 may have served as a storage place for the statuette when it was not on display. The excavators say that this supports their opinion that the kouros was an Osiris-like figure who, associated with the constellation Orion, "rose" from the dead; that is, was brought out of the pit in order to mark the end of the agricultural year, at the harvest, the time when Orion rose heliacally each year (MacGillivray et al. 2000: 169). The argument is that the harvest and winnowing period, signalled by Orion's rising, marked the end of the year in Hesiod's time (ca 700 BC). The Cretans, it is suggested, may have done as the later Athenians did and begun the first month of the year with the first new moon after the summer solstice and had to complete the harvest by the next month when Orion and Sirius culminated and Arcturus was rising at dawn. Scenes of bull-leaping in Minoan art are then interpreted as reflective of the respective positions of the constellations Orion, Taurus, Perseus, and Andromeda, with the last three signifying the hero Theseus rescuing Ariadne from the Minotaur. "It is a very obvious narrative sequence to all who gaze upward at the winter night sky over the northern hemisphere. ... The Greek myth of Theseus, Ariadne and the Minotaur could be nothing more than recollections of the tales created as mnemonic devices to remember the yearly procession of the constellations", with the reappearance of Orion before sunrise in mid-summer possibly marking the completion of the old year (MacGillivray et al. 2000: 128). (Fig. 2 presents a possible visualisation of this interpretation, with a bull-leaping fresco from Knossos, slightly adapted, overlaid on the constellations noted by MacGillivray.)

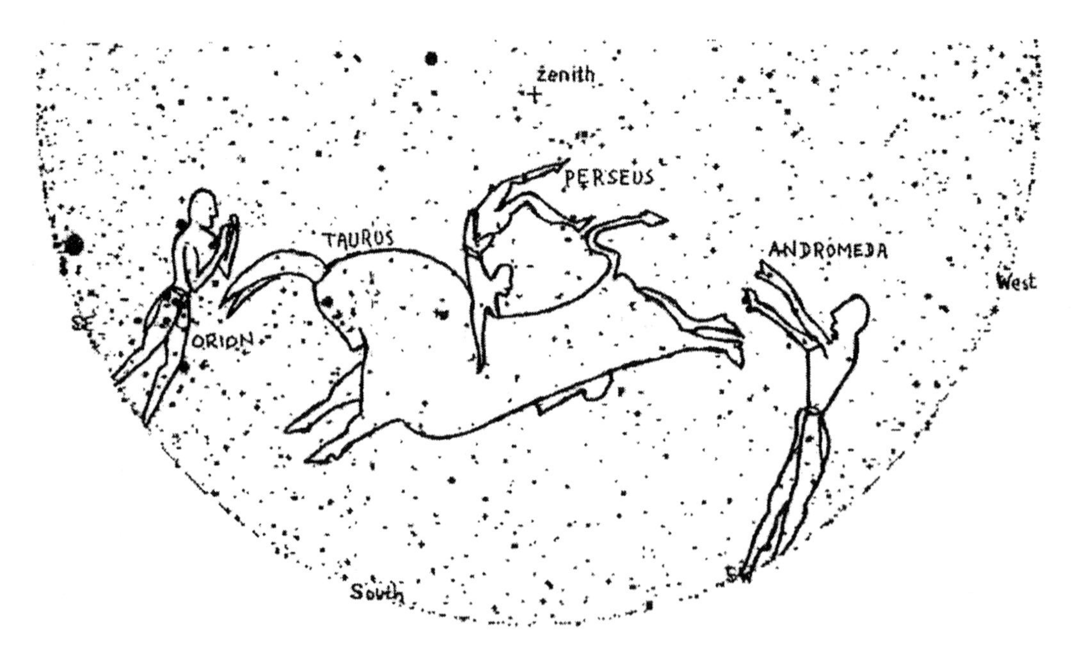

Fig. 2. Star chart of Orion, Taurus, Perseus, and Andromeda, 1500 BC, latitude 35°16′ N. After Voyager II (Carina Software, San Leandro) overlaid with figures based on the Taureador Fresco from the palace at Knossos.

Let us consider these arguments. To help identify the kouros as a deity, the excavators pointed to a seal from Kydonia on which a young man appears in a similar pose (MacGillivray et al. 2000: 128-129). This Late Minoan seal (Fig. 3) shows a man standing above a pair of horns of consecration (Betts 1965: 203-206). He is flanked by two fantastic beasts—a winged goat and a 'genius' or demon—that offer a jug to the man. The composition of the seal, with the man placed higher than his companions, suggests that he is the dominant figure. Hallager states that he is "considered by most scholars to be a god" (Hallager 1985: 24).

Fig. 3. Lentoid seal from Kydonia (after Nilsson 1950: fig. 56).

But when we come to the excavators' astronomical and calendrical interpretations, there are serious problems of methodology. Three different calendrical traditions are inexplicably intertwined: (1) that of Hesiod's farmers of about 700 BC with (2) that of the Athenians as we know it from the fifth century BC, neither of which was bound to the other and neither of which need have had anything at all to do with the third tradition, (3) the Cretan calendar around 1500 BC. For Hesiod, whose "main concern is the cereal crop" (West 1978: 52) and for whom the harvest was certainly important, nevertheless the year began and ended with ploughing, in late October, when the Pleiades, the Hyades (i.e., Taurus), and Orion "set" (West 1978: 125 [*Works and Days* 614-617]).

The Athenian year in the fifth century BC began at a different time, with the first new moon after the summer solstice. However this was the time not of harvest as such, but of threshing the harvest. Furthermore, there is no reason to suppose that the Cretans started their year as the later Athenians did. If we look elsewhere, the Minoans' distant neighbours, the Egyptians, started the new year with the heliacal rise of Sirius, which in 1500 BC was on 1 July in modern terms. This was certainly after the solstice (by a week), but it was a relatively fixed date in contrast to the Athenians' movable lunisolar New Year's day. The Babylonians, on the other hand, also started their new year with a first new moon, like the Athenians, but it was after the spring equinox (i.e., after 21 March).

What we can say did happen in 1500 BC in Crete is that Orion began rising heliacally (measuring from γ Orionis) at Palaikastro from about 11 June in modern terms, almost two weeks *before* the solstice on 24 June. If this marked the end of the old year and start of the new, then it was a different starting point than was used elsewhere at that time in or near the Mediterranean, or even later in the Greek world. This does not make it impossible, but better evidence is required to substantiate the hypothesis than what has been presented so far.

As for the configurations of the constellations identified by the excavators in the description of the bull-leaping scenes, there is no evidence to suppose that the Minoan Cretans grouped the stars into figures in this way, attractive though the configurations may appear when viewed against the 'Bull Leaping' images. The Babylonians saw a bull's head in Taurus, but for Orion they had 'The True Shepherd', for Perseus 'The Old Man', and for most of Andromeda 'The Stag' (α Andromedae being part of 'The Field', with Pegasus). There is no obvious narrative interconnection between these Babylonian equivalents, nor do the figures match up with the 'Bull Leaping' types. The Egyptians saw Osiris in Sah (= parts of Orion); the 'Thigh of Seth' in Meskhetiu (= the Big Dipper within the constellation of Ursa Major), representing it variously as the leg of an animal, or a bull-headed object, or a whole bull; and Isis in Sothis (= Sirius). Beyond these, nothing else is certain. So we do not know what the Egyptians imagined in the stars around Orion (Hannah 1997: 33-39). The fact that they found their Bull in the sky in quite a different place from the Babylonians and Greeks should warn us against the excavators' interpretation: it is in fact not "very obvious", to judge from ancient sources.

The excavators suggest that the minute gold discs set in Egyptian blue pigment found with the fragments of the statuette represented "the starry sky on which he [the kouros] walked" (MacGillivray et al. 2000: 166), an interpretation without parallel. Instead, the golden discs might have been symbols of the *sun*. The iconography on the Kydonia seal supports this, as the genius and the sacral horns are associated with the Egyptian sun god.

A gold ring from Tiryns, roughly contemporary with the Kydonia seal and the kouros, shows a procession of demons bringing liquid offerings in jugs to a seated goddess. Higgins asserts "the Cretan demon was a hybrid monster derived from the Egyptian fertility goddess Ta-urt" (Higgins 1967: 102; Lurker 1980: 119, 120; Weingarten 1991). Ta-urt (or Tawaret) was the Hippopotamus goddess who was associated with, among other things, the sustenance and protection of the dead Pharaoh and, in the Middle Kingdom (ca 2122/2080-1801/1759 BC, Kitchen 1987: 49). She protected the sun god each day as he battled the forces of darkness and was reborn with the dawn (Weingarten 1991: 14). The way in which the genii on the ring from Tiryns hold their vessels, at the base and handle, is an example of what is known as the "Egyptian offering gesture" (Weingarten 1991: 7). These genii, therefore, may be paying homage to a fertility goddess who is also symbolised by the grain above the procession scene. If she is a goddess (Higgins 1967: 188), then the man on the Kydonia seal, because of his central position and the pose of the demon, must also be a god. So too, then, may be the Palaikastro kouros.

The horns of consecration on the Kydonia seal are likely to be a combination of the Egyptian hieroglyph for horns (wpt: -) and that of the valley between two mountains (dw: :) (Newberry 1909: 24-31). Powell states that the Egyptians used the horizon symbol (3ht: ;), "to represent the solar god's place of rising" (Powell 1977: 72; Bernal 1991: 164, 568 n. 46). In other words, the sun was thought to appear between the mountains of the horizon.

Furthermore, in Egyptian iconography the goddess Hathor was depicted as a woman, or as a cow, with horns on her head or headdress. Often there was a solar disc between these horns (Daumas 1977: 1024; Clark 1959: 16, 89). She was a celestial deity, one of whose functions was to nourish both gods and people. It was she who was also believed to carry the solar disk and rejuvenate it daily, as she was its mother (Faulkner 1969: 132), a task also carried out by Ta-urt. With other bovine imagery in Minoan symbolism it is possible that the horns of consecration represented a deity not unlike Hathor, but not necessarily Hathor as such.

On the Kydonia seal, then, it seems that the man is protected by two goddesses: Hathor represented by the horns of consecration and Ta-urt as the Minoan genius, both goddesses being associated with guiding the sun god to the dawn. So there is the possibility that the seal shows the sun god personified, rising, with the protective horns of his mother Hathor below him and Ta-urt beside him.

To return to the Palaikastro kouros, Egyptian influence on the construction, size, materials, and striding pose seems likely. But the kouros is not all Egyptian. The style of the hair is typically Minoan (MacGillivray et al. 2000: 134-135), and the pose of the arms, while encountered in Egyptian art, is more analogous to that of the figurines from Petsophas and Traostalos. So the kouros represents a fusion of elements from both Egypt and Minoan Crete.

The excavators propose that "he was the personification of the youthful male god who arrived from the underworld to herald the beginning of the Harvest: Diktaian Zeus, associated with the Egyptian god Osiris and immortalized as Orion" (MacGillivray et al. 2000: 169.) The evidence suggests instead, however, that this remarkable figure may be a Minoan representation of a solar god whose iconography had its origins in Egypt.

Acknowledgements

Marina Moss would like to thank the University of Otago Research Committee for a University of Otago Bridging Grant that provided support towards the preparation of this paper.

*Classics Department, University of Otago, P.O. Box 56, Dunedin, New Zealand,
robert.hannah@stonebow.otago.ac.nz.
**Classics Department, University of Otago, P.O. Box 56, Dunedin, New Zealand.

References

Bernal, M. 1991. *Black Athena*, vol. 2. New Brunswick.

Betts, J. H. 1965. Notes on a Possible Minoan Forgery. *Annual of the British School at Athens* **60**, 203-206.

Clark, R. T. R. 1959. *Myth and Symbol in Ancient Egypt.* London.

Daumas, F. 1977. Hathor. In *Lexikon der Ägyptologie*, vol 2, ed. W. Helck and O. Eberhard, 1024-1033. Wiesbaden.

Faulkner, R. O. 1969. *The Ancient Egyptian Pyramid Texts*. Oxford.

Hallager, E. 1985. *The Master Impression: A clay sealing from the Greek-Swedish excavations at Kastelli, Khania* (Studies in Mediterranean Archaeology 69). Göteborg.

Hannah, R. 1997. The Tethering of Meskhetiu. *Göttingen Miszellen* **160**, 33-39.

Henriksson, G. and Blomberg, M. 1996. Evidence for Minoan astronomical observations from the peak sanctuaries on Petsophas and Traostalos. *Opuscula Atheniensia* **21**, 99-114.

Higgins, R. 1967. *Minoan and Mycenaean art*. London.

Kitchen, K. A. 1987. The basics of Egyptian chronology in relation to the Bronze Age. In *High, Middle or Low? Acts of an international colloquium on absolute chronology held at the University of Gothenburg, 20th-22nd August, 1987*, ed. by P. Åström, 37-55. Göteborg.

Lurker, M. 1980. *The gods and symbols of Ancient Egypt*. London.

MacGillivray, J. A., Driessen, J. M., and Sackett, L. H. 2000. *The Palaikastro kouros: A Minoan chryselephantine statuette and its Aegean Bronze Age context*. London.

MacGillivray, J. A., Sackett, L. H., Driessen, J., Farnoux A., and Smyth, D. 1991. Excavations at Palaikastro, 1990. *Annual of the British School at Athens* **86**, 121-147.

Newberry, P. E. 1909. Two cults of the Old Kingdom. *Liverpool Annals of Archaeology and Anthropology* **1**, 24-31.

Nilsson, M. 1950. *The Minoan/Mycenaean religion and its survival in Greek religion*, 2nd rev. ed. Lund.

Perlman, P. J. 1995. Invocatio and imprecatio: The hymn to the Greatest Kouros from Palaikastro and the oath in ancient Crete. *Journal of Hellenic Studies* **115**, 161-167.

Powell, B. 1977. The significance of the so-called "Horns of Consecration". *Kadmos* **16**, 70-82.

Weingarten, J. 1991. *The transformation of Egyptian Taweret into the Minoan Genius: A study in cultural transmission in the Middle Bronze Age*. (Studies in Mediterranean Archaeology 88). Partille.

West, M. L. (ed.). 1978. *Hesiod: Works and days*. Oxford.

The *Phaenomena* of Aratus, Orphism, and ancient astronomy

Sergey Zhitomirsky*

Abstract

This paper explores the possible roots of the ancient Greek astronomical poem *Phaenomena* by Aratus (3rd century BC). While the immediate source of the text is a book by Eudoxus of Cnidus (4th century BC), there is some evidence which points to a much longer history. Analysis of the astronomical data contained in the *Phaenomena* (simultaneous risings and settings of the constellations, places of intersection between the constellations and the celestial circles) suggests a dating as early as the boundary of the 3rd and 2nd millennia BC. This means a tradition going on from far antiquity to the times of Eudoxus and Aratus. We suggest that it is the tradition of Orphic mythology, which has some very archaic features characteristic of the Indo-Iranian mythological system, most notably the idea that the universe was created by Ageless Time and the myth of the World Egg with the sky as its shell. This latter concept must have been most important for the formation of the notion of the celestial sphere. It is well known that Pythagoras and his disciples were connected with Orphism. One of the disciples, Architas, was the teacher of Eudoxus whose book served as the source for the *Phaenomena*. Thus the basic ideas of spherical astronomy may have been handed down to them from the founders of ancient science

It is well known that the astronomical poem *Phaenomena* was created by the Greek poet Aratus (ca 310-245 BC), who lived at the court of the Macedonian king Antigonos Gonatos. The king asked him to render in verse the book written by the famous ancient scholar Eudoxus of Cnidus (ca 403-350 BC). But this poem turned out to be only the last phase of the very long and complex history of the text. The description of the starry sky, which is the main subject of the poem, can be objectively dated using astronomical methods. The investigations showed that the sky described in the poem was observed not later than in the second half of the 3rd millennium BC (Zhitomirsky 1999). Consequently, the work of Eudoxus is based on the data of very old astronomical observations, and there was a tradition that delivered to him this information.

How could the scholar of Classical Greece get the results of observations that had been carried out fifteen centuries earlier? Some conjecture that Eudoxus described the sky according to an old celestial globe that had accidentally survived. It is hard to imagine, however, that an article of that kind could have been preserved for such a long time; it is even more doubtful that a celestial globe had existed 1500 years before Eudoxus. We may also remember that Cicero thought Eudoxus to be the inventor of the globe with constellations (Keyes 1943: 41 [*De re publica* 1.14.22]). The most probable supposition is that only a text, which might have come originally from en oral tradition, could fix and preserve such specific information for fifteen centuries.

The description of the star movements in the *Phaenomena* as a whole agrees with modern conceptions of spherical astronomy; Aratus mentions the axis of the universe, the celestial equator, the tropics, and the zodiac circle that touches them. It seems to be obvious that the old data used by Eudoxus were related to names and mutual positions of constellations and that the conceptions of sky geometry are his own. But the detailed investigation showed that this was not the case (Zhitomirsky 1999).

The first method of dating used by Maunder (1909) and Crommelin (1923) was to study the part of the celestial sphere that was unknown to Aratus. It was shown that the centre of the 'invisible area' (obviously due to the precession of the earth's axis) is noticeably displaced compared with the South Pole of the earth in the time of Eudoxus. According to Maunder's dating (based on the magnitude of this displacement), the first, primary source of the poem was created approximately in 2500 BC. The results of Crommelin were similar. Therefore it was proved from this that Aratus' description of the celestial sphere as a whole is of extreme antiquity. Ovenden (1966) came to a similar conclusion, having analysed the data of simultaneous risings and settings of the constellations. According to his results the observations in the poem went back to 2600 BC ± 800 years. The work of Ovenden confirmed the antiquity of Eudoxus' sources and Maunder's dating. Finally, Roy (1984) examined one more series of data in the *Phaenomena*, the places of intersections between constellations and the celestial circles—the tropics and equator. His result, boundary of the 3rd and 2nd millennia BC, was tested by more precise reiteration performed by the present author (1999). These works proved that knowledge of the celestial circles was the essential content of the original text. Moreover, analysis of the poem showed some archaic features that do not correspond to the views of Eudoxus as known from other sources. It seems that the scholar used an ancient text without changing it notably.

So it was proved that there was an astronomical text going back to extreme antiquity that contained a detailed description of the starry sky and developed conceptions of celestial movements. The text was applied practically, judging by the description of simultaneous risings and settings of constellations in different seasons; it was necessary for a calendar. Moreover, the list of equatorial constellations was useful for orientation; they marked east-west directions near the horizon. Finally, this description helped in the remembering of the mutual positions of constellations.

What tradition could serve as a bridge between the time of Eudoxus and the epoch Stonehenge? We may assume with a certain extent of confidence that the role of mediator was played by Orphism, one of the most famous religious trends in ancient Greece. It was formed about the 6th century BC, in connection with the cult of Dionysus. But it seemingly traces its roots back to great antiquity.

The main sources of Orphic doctrine are the preserved fragments of the theogonies and Orphic hymns. Modern scholarship is aware of fragments of four theogonies (Kern 1972; Lebedev 1989; West 1983), which survived in neoplatonic works. The most informative is the so-called *Rhapsodic Theogony* (the sacred story in 24 rhapsodies) and the *Theogony according to Hieronymus and Hellanicus*; both seemingly go back to the 5th century BC. These theogonies let us know of Orphic cosmogony, but the hymns give us more information about characteristic features of the structure of the universe (Taho-Godi 1980: 149-207). The collection of 87 Orphic hymns goes back probably to the 2nd century BC and, judging by it's style, was written by one author, or rather editor, because it is presupposed that any sacred text follows ancient traditions. The hymns include summons of gods and requests to them. Information on astronomic views can be found in symbolic form in the hymns to Uranus, Gaea, the Stars, Apollo, the Mother of gods, and some others.

The antiquity of Orphic views is proved by the presence of several features that are characteristic for Indo-Iranian mythological systems (Braginsky 1980; 560-565):

1) Conception that the universe was created by Ageless Time; Orphics name it *Khronos Ageros Apeiros*, Iranians, *Zurvan Akarna.*
2) Myth of the World Egg, which is common to Indians, Iranians, and Orphics. The latter say that the god-demiurge Phanes was born from the Egg; Indians call this god Brahma. One of the epithets of Phanes, Photogonos (the first-born), is the direct translation of the Indian *prajapati*, which is one of Brahma's names.
3) Transmigration of souls, also common to Indians, Iranians, and Orphics.

There are some other features proving the antiquity of Orphic roots, e.g. the fact that the theogonical myths of the Orphics are complex in comparison with traditional Greek ones (those of Hesiod). Some parallels between the heroes of these myths (Ormazd-Phanes, Ahriman-Night) can also be mentioned in witness of similarity. Our sources are rather late, but nevertheless we have reasons to believe that they reflect very ancient conceptions.

Thus the *Phaenomena* of Aratus begins with the introduction where the poet identifies the cosmos with Zeus (Kidd 1997; 73 [Aratus 1-15]). Scholars usually ascribe this identification to stoic influence, which is natural as Aratus was the disciple of Zeno of Citium, the founder of the stoic school. But there is an earlier identification of Zeus and the cosmos. We find it in the *Rhapsodic Theogony* where Zeus, following the advice of Night, swallows up the universe with all gods and, so to say, recreates it. The last hymn to Zeus (Kern 1922: 199-201 [*Orphica fragmenta* 167]; Lebedev 1989: 54) is very close to the Aratus passage. Aratus may have followed Zeno, who could have been under the influence of Orphico-Pythagorean conceptions. But it is also possible that the introduction belonged to the original text and was accidentally similar to the stoic views of the poet.

It is important to mention, however, that Aratus constantly uses the name of Zeus for starry sky. This deification of sky, characteristic for early stoicism and Orphism, may have traced back to the primary stage of the text; it might be a religious percept or a 'portrait of deity' in a sense (it is known that texts of this kind can be translated into other languages and preserved for a long time).

We can try to date Orphic mythology by comparing it with the conception of the system of the universe in Aratus' poem, which has received objective astronomic dating (not earlier than 2000 BC). This dating is also substantiated by the archaeological evidence: finds in the ruins of the Minoan shrines on Petsophas and Traostalos in Crete. The clay figurines found there seemingly symbolise constellations that largely correspond to those described by Aratus (P. Blomberg 2000, Henriksson & Blomberg 1996, 2000).

It has already been said that the philosophical views on the universe of the Orphics agree with those of Aratus. In both cases we find deification of the universe and belief in the Law that rules it. More detailed parallels between Orphics and Aratus can be found in their astronomical conceptions. The hymns and fragments of the theogonies give us enough material to learn Orphic views on astronomy. First, we can find the conception of the spherical sky that encircles the earth in Orphic hymns; this idea was not trivial. In any case it is absent in the most developed Mediterranean cultures of the epoch, Sumero-Accadian and Egyptian. It may be suggested that this conception is connected with the idea of the World Egg. The Indian, Iranian, and Orphic mythologies place the universe inside a metal egg; its shell is the celestial sphere. The Orphic myth expounded in the *Theogony according to Hieronymus and Hellanicus* is close to the Indian; in both cases the Egg is born in pristine waters and floats there (Toporov 1982; 681). It is important that the Egg arises as a result of rotation and revolves. The *Rhapsodic Theogony* also describes the rotating Egg.

The conception of the solid spherical sky is reflected in the hymns to the Night (Quandt 1941: 4 [*Orphei Hymni* 3]; Taho-Godi 1980), to Uranus (Quandt 1941: 5 [*Orphei Hymni* 4]; Taho-Godi 1980), and to Gaea (Quandt 1941: 22 [*Orphei Hymni* 26]; Taho-Godi 1980). Two of them (3 and 4) include the mention of the Universal Law—Night is ruled by *Ananke* (necessity). In the hymn to the Mother of gods (Quandt 1941: 22 [*Orphei Hymni* 27]; Taho-Godi 1980), we find the axis of the universe, and in the hymn to Apollo (Quandt 1941: 27 [*Orphei Hymni* 34]; Taho-Godi 1980), the axis and, seemingly, the celestial circles (they are symbolised by the strings of the lyre that let Apollo control the change of seasons). The Orphics thought the earth to be flat. It is directly said in the hymn to Oceanus, "You wash the circle of earth, having limited it to yourself" (Quandt 1941: 55 [*Orphei Hymni* 83.3]; Taho-Godi 1980). The ring of Oceanus that surrounds the earth is also mentioned in the *Rhapsodic Theogony* (Kern 1922: 178 [*Orphica fragmenta* 115]; Lebedev 1989: 51), and some details in the *Theogony according to Hieronymus and Hellanicus* presuppose the existence of the universal directions up and down: the Egg was born in the depth and rose up; Phanes is sitting on the top of the earth (Kern 1922; 132-133 [*Orphica fragmenta* 55]; Lebedev 1989: 62).

The main elements of this conception agree with Aratus' image of the universe. As has already been mentioned, Roy and the present author showed that the idea of celestial circles traces back to the original of the *Phaenomena*, and consequently the spherical sky and other conceptions of spherical astronomy are of the same age. Aratus says nothing about the form of the earth, but there are hints that it is considered flat (it is well-known that Eudoxus thought the earth to be ball-shaped). The universe of the *Phaenomena* is divided vertically into the bright upper 'aerial' part and dark 'underwater' part—Oceanus, described as a dark abyss that envelops the earth from below. It hides parts of celestial circles and heavenly bodies after their settings 'in the waves' and in the darkness 'of long (i.e. eternal) night'. The fact that the earth in this conception is flat is proved by terms for the horizon, and the description of the tropics as inclined to it. Aratus calls the horizon 'ocean', 'the edge of the universe', and 'the earth'. E.g. he writes of the celestial equator, "The third circle is rotated, being divided by the earth into two parts" (Kidd 1997; 111 [Aratus 511-512]). It is hard to imagine how the celestial sphere could be divided into two parts by a ball-shaped figure. Thus the Orphic texts are describing the spherical cosmos with the inhabited area in the centre of the flat round earth. The latter is surrounded by the celestial sphere that eternally rotates round its indissoluble axis. This picture agrees in general with the cosmos described by Aratus who says that the celestial sphere is half-immersed in 'dark Oceanus' and also included the flat earth. It is necessary to note that this conception of the universe is an excellent model for spherical astronomy.

The important role of Orphic conceptions in the formation of crucial ideas of ancient astronomy was emphasised more than 30 years ago by Chassappis in Greece (1967) and Veselovskij in Russia (1982). Since then a number of investigations were carried out that proved many of their pioneer suppositions. We have reasons to think that Orphism had introduced to Greek science two non-evident basic ideas. The first one is the conception of the spherical sky surrounding the earth; the second one is the belief in the Law that confirms universal concord including mathematical harmony.

Lebedev (1978) came out with a persuasive suggestion that Anaximander thought the Ageless Infinite Time (Χρόνος Αγήρως Ἄπειρος) of the Orphics to be the origin of everything. The universe of the philosopher is subjected to strict mathematical ratios. There is no direct mention of the sphere of stars, but there are reasons to suppose that this sphere was implied. The system of the cylindrical earth suspended in the centre of the sphere of stars basically agrees with systems of the universe in Orphism and the *Phaenomena*. There is one more parallel between the views of the Orphics and Anaximander. The philosopher thought that the planets, the moon, and the sun were holes in the dark circles surrounding the earth and that they were filled with fire. In the Orphic hymn to the stars, "sevenfold belts" are mentioned (Quandt 1941; 7 [*Orphei Hymni* 7.8]). We may suppose that the matter concerns the circles of heavenly bodies.

It is well known that Pythagoras and his disciples were connected with Orphism (thus the 'Pythagorean Cecrops' is called the author of the *Rhapsodic Theogony*). Along with philosophy, politics, and religion, Pythagoras was interested in mathematics. His search for geometrical harmony in the universe had seemingly brought him to the conclusion of the spherical earth, which was an astronomical hypothesis of great importance. Plato was a friend of the Pythagorean Architas of Tarentum, and Eudoxus, who created the first model of astronomical motions, was his disciple. We would not belittle the merits of the scholars who founded ancient science, saying that the basic ideas of spherical astronomy were delivered to them in ready form from unknown priests-astronomers of the times of Stonehenge by means of Orphic religions.

*Apt. 143, 3-1 Obrucheva St. 117421 Moscow, Russia, sergey@panina.dnttm.ru.

References

Blomberg, P. 2000. An astronomical interpretation of finds from Minoan Crete. In *Oxford VI and SEAC 99: Astronomy and cultural diversity*, ed. C. Esteban and J. A. Belmonte, 311-318. Santa Cruz de Tenerife.

Braginsky, I. S. 1980. Iranian Mythology. In *Myths of the peoples of the world* (in Russian), vol. 1, 560-565. Moscow.

Chassappis, C. S. 1967. *Greek Astronomy of the 2nd Millenium BC* (in Greek). Athens.

Crommelin, A. C. D. 1923. *Splendour of the heavens*, vol. 2. London.

Henriksson, G. and Blomberg, M. 1996. Evidence for Minoan astronomical observations from the peak sanctuaries on Petsophas and Traostalos. *Opuscula Atheniensia* **21**, 99-114.

Henriksson, G. and Blomberg, M. 1997-1998. Petsophas and the summer solstice. *Opuscula Atheniensia* **22-23**, 147-151.

Kern, O. (ed.) 1922. *Orphica fragmenta*. Berlin.

Keyes, C. L. (tr.) 1943. *Cicero: De re publica* (Loeb Classical Library 213). Cambridge MA, London.

Kidd, D. (tr.) 1997, *Aratus: Phaenomena* (Cambridge Classical Texts and Commentaries 34) Cambridge MA, London.

Lebedev A. V. 1989. *Fragments by early Greek philosophers* (in Russian). Moscow.

Lebedev, A. V. 1978. *Τὸ Ἄπειρον: Not Anaximander, but Plato and Artistotle* (in Russian) (Studies in Ancient History 1,2). Moscow.

Maunder, E. W. 1909. *The astronomy of the Bible*. London

Ovenden, M. W. 1966. The origin of the constellations. *Philosophical Journal* **3**, 1-18.

Quandt, W. (ed.) 1941. *Orphei Hymni*. Berlin

Roy, A. 1984. The origin of the constellations. *Vistas in Astronomy* **27**, 176-185.

Taho-Godi, A. A. 1980. *Ancient Hymns* (in Russian). Moscow.

Toporov, V. N. 1982. The World Egg (in Russian). In *Myths of the peoples of the world*, vol. 2, 681. Moscow.

Veselovskij, I. N., 1982. Astronomy of the Ophics (in Russian). In *Problems in the history of natural science and engineering*, vol. 2, 120-124. Moscow.

West. M. L. 1983. *The Orphic poems*. Oxford.

Zhitomirsky, S. 1999. Aratus' *Phaenomena*: Dating and analyzing its primary source. *Astronomical and Astrophysical Transactions* **17**, 483-500.

The role of celestial routes of nocturnally migrating birds in the calendrics and cosmovisions of ancient peoples

Izold Pustylnik*

Abstract

We summarize the scientific evidence favouring the theory of a 'stellar compass' that is used by several species of birds during their seasonal migrations and then discuss some of the theory's plausible implications in the cultural context of archaeoastronomy. From time immemorial man venerated the birds, but also was compelled to kill them. We put forward an idea that this dichotomy, accompanied by an intuitive complex of guilt on the part of our distant ancestors, may be reflected in the sublime forms of bird totems and rituals associated with birds. We suggest that through direct experience of the habits of birds primitive people first became aware of the everlasting succession of events in an ambient world and, once perceiving it, these people incorporated birds into their system of complex rituals, beliefs, and cosmovision. This view naturally explains why birds in general and waterfowls in particular, the ubiquitous inhabitants of all three elements—air, water, and land—are invariably encountered in ancient myths and legends of very different cultures all over the globe and have always stirred the creative dreams of man.

There is growing evidence that zodiacal constellations were fashioned prior to the introduction of written language and long before the dawn of astronomy in ancient Babylon and Greece. It has been argued that an intuitive perception of stellar patterns could have been an essential ingredient of an inborn pattern-recognition system of *homo sapiens* and hence could lie at the heart of a complicated guidance system for nomadic tribes of hunters and gatherers during the Palaeolithic era (Alerstam 1990). Nature shows us that human beings are far from being unique in their capacity to identify stellar patterns and to use them as part of a reference system. From prehistoric times man has revered the birds, but at the same time partially domesticated them. Certainly from ancient Egypt up to modern times he employed some species as messengers (Keeton 1974).

Now numerous contemporary field and laboratory investigations unequivocally suggest that several species of nocturnally migrating birds (e.g. the indigo buntings, blackpoll warblers, red-backed shrikes, and several others) use circumpolar stellar constellations as an essential element of their sophisticated navigation system in seasonal migrating journeys covering sometimes up to four, even six thousand miles (Emlen 1975). In the first part of this contribution we summarize briefly the available evidence for the use of a 'stellar compass' by migrating birds (following basically Emlen 1975 and Alerstam 1990) and thereafter we proceed to its probable implications discussed in the cultural context of archaeoastronomy. In principle two alternative modes of navigation by stars are conceivable:

1) Nocturnally migrating birds may locate some star and guide themselves by flying at a particular angle relative to it. However, due to the earth's rotation, changes occur in the azimuth of a chosen star during the night. Consequently, the bird should alter the angle of its flight to compensate for the diurnal motion of a specific star.
2) Another way of navigating is to use the fixed-star patterns and their relationships to each other. For instance, by the characteristic arrangements of the stars in the constellation of Ursa Major we readily locate the Polar star and hence geographic north. Although the Ursa Major pattern also moves across the sky, its shape remains intact and it retains its relationship to Polaris. So as long as Ursa Major or any other circumpolar stellar pattern is visible on the firmament, we can determine north without knowing the time in the night, season, or the geographic location (Fig. 1).

Obviously the system must possess a considerable degree of redundancy. Even if the sky is partially overcast, an experienced bird can still identify some single constellation. It is needless to emphasize that an exact determination of north and a flying route is of a vital importance for birds that show a remarkable fidelity in choosing breeding sites and winter grounds.

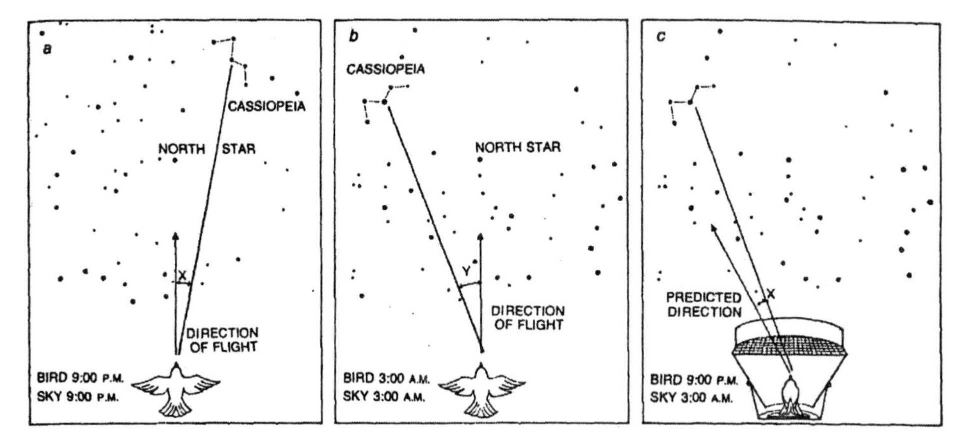

Fig. 1a-c. Stellar orientation hypothesis 1 proposes that the indigo bunting guides itself by flying at an angle to a particular star or group of stars. Since the positions of the stars change throughout the night, the bird would have to use an internal time sense to compensate for the motion of the stars. For example, a bunting going north at 21.00 would fly at an angle X with respect to a critical star (*a*). At 03.00 the bird compensates for the rotation of the stars by flying at angle Y to a critical star (*b*). According to the hypothesis, when a bunting whose physiological time is 21.00 is presented with a 03.00 star pattern in a planetarium, it should compensate in the wrong direction; that is, it should orient at angle X with respect to the critical star instead of at angle Y (*c*).

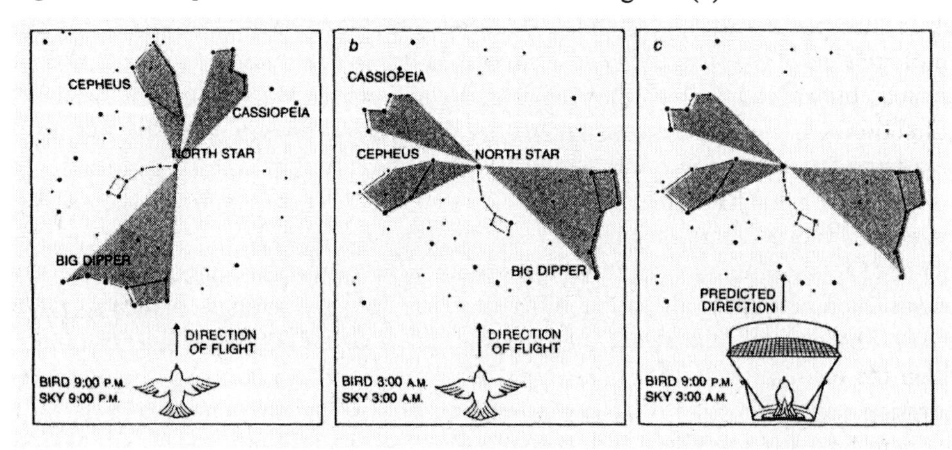

Fig. 1d-f. Stellar orientation hypothesis 2 states that the bunting obtains directional information from the configuration of the stars. The bird can determine a reference direction such as north from fixed geometric relationships of the stars regardless of the time of the night (*a, b*). When the bunting is exposed to a time-shifted sky in a planetarium, there should be no change in its orientation (*c*).

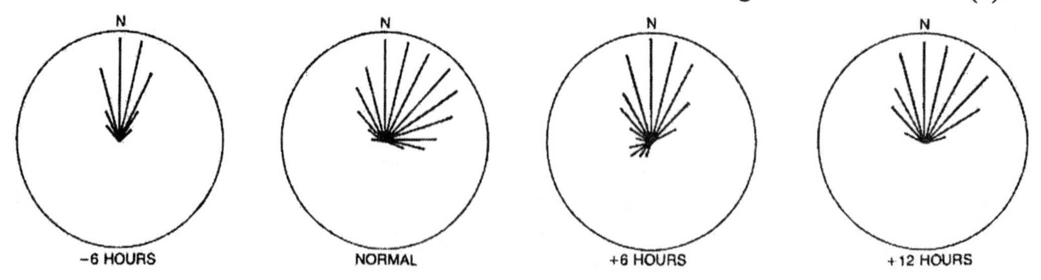

Fig. 1g-j. Results of planetarium tests of the two star-navigation hypotheses show that buntings continue to orient correctly regardless of whether planetarium stars are shifted ahead of or behind the bird's normal physiological time. This indicates that the bunting does not incorporate its biological clock in the star-orientation process and obtains only directional information from the star patterns. Permission to publish courtesy of the artist Mr. Adolph E. Brotman.

These alternative hypotheses have been tested under artificial skies in a planetarium. In Fig. 2 the bird is shown inside a circular cage, which is made of white blotting paper folded to form a funnel (conveniently now called Emlen's funnel by ornithologists) and mounted on a base consisting of an inkpad. As can be seen from this figure, the indigo bunting hops onto the test cage. Black marks on the white side of the cage are footprints left by the bird. *The bird's view is limited to a 140-degree overhead sector of the sky when it hops up.*

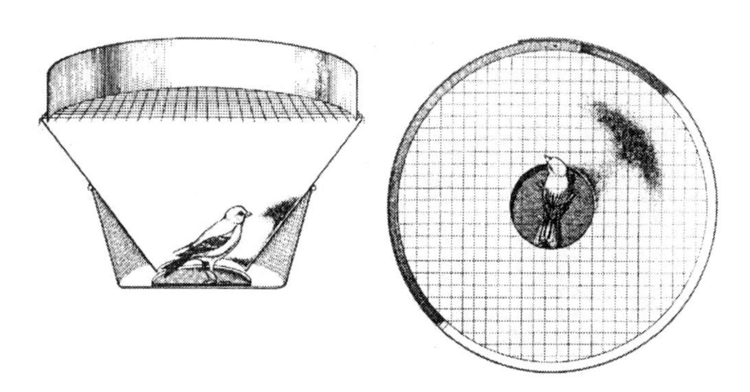

Fig 2. Circular test cage for determining the directional preference of an indigo bunting is shown in cross section and in top view. Funnel portion of the cage is made of white blotting paper. The bunting stands on an ink pad and each time it hops onto the sloping funnel wall it leaves black footprints. The bird's view is limited to a 140° overhead sector of sky when it hops up. Permission to publish courtesy of the artist Mr. Adolph E. Brotman.

Experiments conducted by several groups of investigators in the United States and Germany under artificial skies in a planetarium indicate that the ability to orient does not involve an internal time sense, a 'biological clock' (needed to compensate for the diurnal motion of a guiding star upon a celestial sphere), but rather a characteristic arrangement of stars; i.e. migrating birds determine their 'compass direction' from geometric patterns of stars.

The same northern circumpolar stars are used as the chief stellar guidance system in both spring and autumn. It turns out that the polarity of migration orientation, whether it is directed *towards* or *away* from the northern circumpolar stars, is under hormonal control. Namely, in special field experiments with white-throated sparrows, the orientation behaviour was reversed by introducing hormones such as prolactin, which have a synergetic effect in stimulating the birds' migratory activity (cf. Emelen 1975).

In addition the tests undertaken with young birds under artificial skies also prove the importance of the maturation process in making use of celestial rotation for a directional reference system. This has been demonstrated in the experiments with the European chaffinches that were captured and displaced from their normal fall migration routes. It turns out that the adult birds may correct for the displacement and reach their habitual winter grounds, while nestlings during their first migratory flight do not. In another set of experiments one group of young indigo buntings was kept in captivity until they became four to ten days old without having seen a point source of light. They lived in a windowless room with diffuse fluorescent lighting. At the same time a control group of young birds was prevented from seeing the sun, but they were exposed to the artificial night sky in a planetarium for a limited period early in autumn when, following inborn instinct, the birds are ready to fly away and manifest great activity and uneasiness. Analysis of the angular distribution of the footprints left by the birds in Emlen's funnels revealed random orientation for the first group. Unlike the inexperienced birds the indigo buntings in the control group were able to orient towards the south (Emlen 1975). The experiments described above "are equally important for what they do not explain about migratory roots" (Emlen 1975). The theory of the stellar compass can explain how the migrating birds may select and maintain a given direction. But the system does not provide information about actual geographic location nor does it explain how a bird can recover its course once it has been blown off to the east or west of it. Ornithologists claim that birds have numerous directional cues in their possession, such as the presence of topographical landmarks, the position of the sunset, direction of the magnetic field, etc. (Keeton 1974, Emlen 1975).

For those who may be interested in this fascinating subject, I have compiled a very fragmentary inventory of the different ingredients of the very tricky navigating system applied by migrating birds.

1) Like our predecessors birds migrate in flocks. Birds save flight energy in this way by flying 'wingtip to wingtip'; they can save almost 40% of their lift power compared with a solitary journey in the air. Another way of saving energy resources lies in a soaring-glide flight, enabling birds to use tail wind more efficiently. During this mode of flight (studied by radar monitoring techniques) birds often use vocal signals or utterances whose echo beam, according to one imaginative theory, can serve in itself as a radar device to give the bird the cue as to the wind direction and even the kind of terrain below. Birds, like human beings, probably build up their map by learning landmarks. However, even at the modest altitude of one kilometre, the range of their direct vision—limited by the curvature of our planet—is over one hundred kilometres. Quacking flocks of wild geese have been reported as close to the very top of Mt Everest (Alerstam 1990).
2) Studies of the solar compass of the migrating birds reveal that birds can perceive the polarized solar light, thereby determining the sun's position with reasonable accuracy, even when the latter is hidden behind clouds or has sunk below the horizon (Alerstam 1990).
3) Both field and laboratory experiments with a number of species of nocturnally migrating birds have revealed evidence for geomagnetic disturbance in their orientation. The respective data have been gathered either by using surveillance radar to track the direction of flights of migrants during actual geomagnetic storms or by introducing an induced magnetic anomaly. These data mean that birds can extract directional information from the geomagnetic field (cf. Moore 1977). Although many mysteries pertaining to the birds' magnetic sense remain unresolved, there is convincing evidence that magnetic material is present in the form of small magnetite crystals, found both in the front and the rear part of a bird's head. Indeed the hypothesis has been put forward that man himself might possess a 'sixth sense', even though we are not conscious of it (Alerstam 1990).

Obviously a detailed description of the sensory world of birds is far beyond the scope of this contribution. According to Emlen, "the discovery of a hierarchy of redundant directional cues *makes the search for a single mechanism of migratory orientation obsolete* "(Emlen 1975).

In an article with the provocative title "Of men and moths - can archaeoastronomy be traced back to the animal kingdom" (Schlosser 1997) as well as in the preface to *Sterne und Steine* (Schlosser and Cierny 1997: 13-17), Schlosser attempts to trace the roots of orientation in space by animals. He argues that this sense of orientation is directed by a type of "setting and tuning" found in all living organisms, a biological clock via in-born tidal rhythms that have paced the evolution of our planet for billions of years. Comparing the angular accuracy for both azimuths and elevations achieved by creators of the monuments and graves of antiquity in orienting them towards cardinal points with the precision of orientation in space shown by bees, pigeons, and other species with respect to the sun, he finds a rough numerical accord between these data. Although in broad terms Schlosser's arguments may prove to be correct one day, specific mechanisms standing behind and regulating spatial orientation of various species still remain a virtual *terra incognita.*

Slightly imitating Schlosser's line of thought, we shall attempt now to trace how the distinctive features of the flights of migrating birds might influence the intuitive perceptions of the ambient world by primitive peoples. Certainly the seasonal migrations of birds would not have escaped the attention and imagination of our ancestors. From time immemorial man has venerated the birds, but also was compelled to kill them. *This dichotomy—an ambivalent predisposition and the nascent intuitive complex of guilt of our distant ancestors—may be seen as the root of the sublime form of bird totems and the rituals associated with birds. In particular it seems plausible that it was through the observation of birds that primitive peoples first perceived the everlasting natural succession of events in an ambient world and, once having perceived it, human beings in return incorporated birds into their system of complex rituals, beliefs, and cosmovision.*

We find one of the best-documented testimonies of the ritual importance of birds reflected in Kalevalian cosmology (Gogin and Kirsanov 1999: 243): "The world is created either a. from an egg (golden or spotted) of a water-bird by the traditional threefold scheme (the upper and lower parts of the shell become heaven and *earth* respectively, the spots become *stars* on the revolving sky, which is supported upon the earth by some *pillar* (*axis mundi*), b. from a pinch of soil obtained by a diving water-bird from the bottom of the primordial ocean."

This idea also naturally explains why birds in general, and waterfowls in particular, are invariably encountered all over the globe in ancient myths and legends of very different cultures. After all, birds are ubiquitous, being the only species that is equally at home in all three elements—air, water, and dry land. From Norway to the Urals rock carvings of similarly profiled ornaments repeat themselves with waterfowls playing a central role. In rock myths concerning the structure of the universe, waterfowls occupy the *medium position*; it is a position of paramount importance for it heralds the arrival of three sets of events: spring—arrival of birds and nestling, summer—the time for moulting and hunting, and autumn—departure of birds (Viktorova 1996). Similar motifs associated with seasonal changes and bird migrations are found in the mythology of Caribbean aboriginal groups (Roe 1993; cf. Lebeuf 1996): "Migratory water birds in flight, carved in ivory were prominent images at Malta."

In both Finno-Ugric and Russian traditions, as well as many other cultures all over the globe, the Milky Way was universally recognized as the way of migratory birds. In his well-known and detailed treatise, Lebeuf (1996: 150) brings together significant information concerning this topic. For example he observes that the Milky Way is the way for the souls of the dead, which they travel through in the shape of birds. Also we find that in the Ukraine, people imagine heaven as the place where birds spend the wintertime. Finally, Lebeuf points out that in Turco-Tatar languages the Milky Way is called the way of birds, the same again among Finns, Estonians, Ostiaks, and Voguls, for whom it is the way of migratory birds.

According to Tultseva (2001), "Such names for the Milky Way as *Goose Way* and *Crane Way* are known in Kaluga region. According to one explanation the Lord put this mark for the heavenly bird to help it find its way to warmer lands."

Discovery of artefacts associated with the bird totems among Ugro-Finnish tribes have been reported by several investigators. Thus Zhuravlev (1996), working at the famous site of Russian wooden architecture in Kizhi, describes the whole cult complex called by him Pegrem-40, located on the bank of Lake Onezhskoje and consisting of 28 stone objects. One of the central monuments is an imposing stone duck with its head directed towards the northeast. According to Zhuravlev the monument dates back to the 3rd and 2nd millennium BC. Kirillin and Pen'kov (1999) describe the result of their study based on six ornamented needle boxes manufactured from tubular bones of birds excavated in the Hayargyz cave in Yakutia and belonging to early Neolithic cultural strata. According to these authors the needle boxes reveal a complicated pattern of incisions that can be interpreted as the primordial lunar and solar calendars used by the ancient hunters.

From prehistoric times up to modern ages man intuitively mimicked nature in most of his inventions. Realization of man's eternal dream to fly is no exception. Are the amazing discoveries of the celestial navigation system of nocturnally migrating birds only part of the slang of modern science, our utilitarian way of expressing what our ancestors *felt* intuitively already long ago, or do we stand on the verge of breathtaking new discoveries, for instance a dialogue with birds via computers?

Up to now, unravelling the secrets of birds in flight, in their manifold interrelations with the human world, remains a truly interdisciplinary enterprise inspiring and attracting physicists, astronomers, archaeologists, historians, zoologists, ethnographers, and artists.

Acknowledgements
The author is indebted to Dr. R. Frank, Dr. M. Blomberg, and Dr. S. Iwaniszewski for fruitful discussions and constructive criticism of this investigation. The author expresses his special gratitude to the anonymous referee who carefully read the manuscript and, in addition to linguistic corrections, offered many helpful suggestions aimed at improving both the style and the elucidation of the main propositions of this research. I am also grateful to Mr. Adolph E. Brotman for permission to use the drawings he made for Emlen's article and to the *Scientific American* for furnishing his address.

Financial support from Grant 4703 of the Estonian Science foundation and from a travel grant from SEAC is gratefully acknowledged.

*Tartu observatory, 61602 Tõravere, Estonia, Izold@aai.ee.

References

Alerstam, T. 1990. *Bird Migration*. Cambridge.

Emlen, S. T. 1975. The stellar-orientation system of a migratory bird, *Scientific American* **233:2**, 102-111.

Gogin, N. D. and Kirsanov, N. O. 1999. Astronomy in Kalevalian epic poetry. In *Actes de la Vème Conférence Annualle de la SEAC, Gdańsk 1977* (Światowit supplement series H: Anthropology 2), ed. A. Lebeuf and M.S. Ziólkowski, 239-250. Warsaw, Gdańsk.

Keeton, W. 1974. The mystery of pigeon homing. *Scientific American* **231**, 96-107.

Kirillin, A. C. and Pen'kov, A. V. 1999. Neolithic needle boxes-calendars from the territory of Yakutia (in Russian). In *Young Archaeology and Ethnography of Siberia and the Far East*. Chita.

Lebeuf, A. 1996. The Milky Way, a part of the souls. In *Astronomical Traditions in Past Cultures: Proceedings of the 1st general meeting of the European society for Astronomy in Culture (SEAC), Smolyan, Bulgaria, 31 August-1 September 1993*, ed. V. Koleva and D. Kolev, 148-161. Sofia.

Moore, F. R. 1977. Geomagnetic disturbance and the orientation of nocturnally migrating birds. *Science* **196**, 681-682.

Roe, P. G. 1993. The Pleiades in comparative perspective: The Waiwai Shirkoimo and the Shipibo Huishmabo. In *Astronomies and Cultures*, ed. C. L. N. Ruggles and N. J. Saunders, 296-328. Niwot.

Schlosser, W. 1997. Of men and moths: Can archaeoastronomy be traced back to the animal kingdom. In *Proceedings of the 4th SEAC Conference: Astronomy in Culture*, ed. C. Jaschek and F. A. Barandela, 255-261. Salamanca.

Schlosser, W. and Cierny, J. 1997. *Sterne und Steine: Eine praktische astronomie der Vorzeit*. Stuttgart.

Tultseva, L. A. 2002. Middle Russian folknames for the Milky Way (in Russian with extensive English abstract). In *Astronomy of Ancient Societies: Proceedings of the European Society for Astronomy (SEAC) associated with the Joint European and National Astronomical Meeting (JENAM), Moscow, May 23-27, 2000*, ed. V. Obridko, T. Potiomkina, A. Lushnikova, and I. Pustylnik, 280-285. Moscow.

Viktorova, V. D. 1996. Mythological world picture of Eneolithic populations in the Trans-Ural forest-mountain regions (in Russian with extensive English abstract). In *Archeaoastronomy: the problem of emergence*, ed. V. V. Volkov, E. N. Kaurov, M. F. Kosarev, T. M. Potemkina, and G. G. Piatych, 23-25. Moscow.

Zhuravlev, A. P. 1996. Cult Complex Pegrem-40 (in Russian). In *Archeaoastronomy: Problems of Formation* ed. V. Volkov, E. Kaurov, M. Kosarev, T. Potemkina, and G. Piatych, 54-56. Moscow.

The power of binding and loosening: Ropes establish the cosmic order

Barbara Rappenglück*

Abstract

Ropes and the use of ropes have left clear traces in the myths and fairy tales of people all over the world. Numerous examples let us surmise that many of these ropes are related to the sky and connected with selected astronomical phenomena. This article categorizes these phenomena more closely, focusing on only a few very representative examples, which give us a clear hint or a well-based probability for an astronomical meaning. In the second place, considerations based on the sciences of symbolics, religion, and ethnology suggest that the astronomical meanings of ropes are connected with their magico-religious ones.

1. Ropes symbolize astronomical phenomena

1.1. Ropes give the basic elements of cosmic order

1.1.1. The rope as axis of the world

According to some myths, in the primordial beginning a rope linked the earth to the sky and thus enabled the formation of the earth (as it is told, e.g., by the Toba Battak of Indonesia: Dixon 1964: 160-161) and the development of human life and culture (Griaule 1980: 50-54). In an example from the Maya the rope linking sky and earth regulated in addition the life of the first beings by providing them with food, like an umbilical cord. But finally "... the rope was cut ... and earth and sky were parted" (Alexander 1964: 153). Here an idea is expressed that occurs again and again in the myths: The rope linking earth and sky constitutes the order of the cosmos by giving stability. The cutting of the rope, endangering the cosmic order by destabilizing it, is in some myths said to have happened in mythical times and meant the end of a paradise-like life; in other myths it is feared that the same will happen again in the future and will mean the end of the world. In both versions the cutting of the rope means a catastrophe, because the rope serves as the axis of the world.

In many myths the connecting rope is not so explicitly characterized as the world axis, but as simply giving access to some special points of the sky:

1.1.2. The rope giving access to the sun, the moon, and the polar point

In myths and fairy-tales all over the world there are endless examples of ropes hanging down from the sky (Hatt 1949: 49, 51-53). They serve as an access to the heavenly regions. In some cases the point where the rope is fixed is specified as being the sun or the moon. Then the rope may be identified with a sunbeam or the moonlight (Luomala 1940: 27; Milbrath 1999: 74). In other cases the rope by which the hero gains access to the sky is explicitly said to be fixed to the polar point (Chuckchee, Pawnee: Hatt 1949: 53, 62).

1.1.3. Ropes bind celestial bodies to the centre of the world

The observation of the stars continuously circling around the polar point may have caused the idea that they are tied to the polar point like animals to a post. E.g. in Kirgisian thought (Harva 1933: 189, more examples 40-42) the three stars of the Little Dipper, which are next to the polar point, are seen as a rope that binds two horses—the two brighter stars of the Little Dipper—to the polar post. Ancient Egypt knew the idea of the Big Dipper being bound to the pole and circling around it (Krupp 1983: 212). According to the Siberian Chuckchee the stars are reindeers tied to the pole by long ropes (Bronsart 1963: 140). While these ropes regulate the stars' movement around this centre, other ropes mentioned in myths seem to represent

1.1.4. The ecliptic, the Milky Way or an encircling 'hoop' of the world

A number of pictures from Mayan codices represent the ecliptic as a cord (Milbrath 1999: 74, 76). Also in Chumash mythology a cord stretching around the world marks the sun's path (Hudson and Underhay 1978: 52). The vision of Er in Plato's *Republic* describes a broad ribbon of light (Shorey 1987: 501 [*Republic* 10.616B-C]), which has been interpreted as either the zodiac or the Milky Way (Dechend and Santillana 1993: 210-211). According to the Pima of Arizona the Milky Way is a cord-like net that connects the borders of earth and sky (Jacobi 1981: 123).

1.1.5. Ropes establish a net of coordinates
With these different examples of ropes important elements of the cosmic order are established, giving coordinates for orientation in place and time, like a system for surveying. The Mayan 'Popol Vuh' (preface of the story-teller) describes the mythical process of surveying earth and sky by means of ropes and thus illustrates their function for establishing the cosmic order. In a text from the Edda (Häny 1978: 211, Erstes Lied von Helgi) the act of surveying concerns the whole cosmic order too, because the threads which are used for surveying the kingdom of a future ruler are fastened in the centre of the moon's hall and in the east and the west. In the myths ropes are not only used to establish cosmic order but also for the

1.2. Manipulation of cosmic movements
1.2.1. Celestial bodies kept on the lead to determine their way
According to a Nigerian myth (Wilson-Haffenden 1927-1928: 144) long chains are attached to the sun and moon, spanning the entire sky. The western and eastern ends of these chains being pulled and released, the sun and moon are enabled to move across the sky. Another African myth mentions a hero who caught the sun, the moon, Orion, and the Pleiades and tied them to a place in the village of the sky god. Then the sky god taught the tied celestial bodies how to move around his village and the sky (Diederichs 1998: 53-55). While these are myths concerning various celestial bodies and the manipulation of their courses by tying them, there also exists a special sort of myth describing the manipulation of the sun's course by snaring the sun.
1.2.2. The ensnared sun: manipulation of the sun's course
The motive of catching the sun in a snare is a well-known element of myths from Oceania, Africa, and North America (Luomala 1940). Two main types of sun-snaring stories can be clearly distinguished. The mainly North American type often consists of the following elements: The sun burnt the hero's cloak. To punish the sun he set a snare, which was made of his sister's hair. The sun got trapped, darkness covered the earth and threatened all life on earth, but the hero failed to free the sun. At last an animal succeeded in releasing the sun. In this type of story the motivation for snaring the sun is to punish the sun for its bad behaviour (Luomala 1940; 46, similar in Africa). In the second main variant the intention of snaring the sun is to slow down its speed and thus to have longer days. E.g. (Jakubassa 1998: 33-37): The Polynesian mythical hero Maui noticed that the days were too short because the sun moved too fast, and darkness threatened the possibilities of having a normal life. Therefore Maui caught the sun in a snare, which was made of a woman's hair. The captured sun was only released when he promised to move more slowly so that the days would become longer. The two main types of the myth have a common element: The sun is caught in a snare and by this action the sun's course is either regulated for the first time or threatened in its regularity. In each case the action of snaring the sun means a manipulation of the sun's course and thus a manipulation of the cosmic order.

The sun-snaring motive of the myths reminds us of magical intentions that were connected by some people with the practice of playing string figures (e.g. the 'cat's cradle' game). Different arctic people knew of a taboo against playing string figures shortly after the winter solstice, because "otherwise the sun might get tangled up in the strings and keep falling down, making its progress much slower" (MacDonald 1998: 117). The sun is seen as a woman whose leggings are torn and whose legs are covered with scars and wounds "... as a result of people continuing to play string-figure games ... after the Sun was back on the horizon" (MacDonald 1998: 126).

Some authors have suggested that the North American type of the sun-snaring motive symbolizes the disappearance of the sun in the polar night, the snare being interpreted as the diminishing diurnal arc (e.g. Hirsch 1965: 86). The cited string-figure practices, coming from the same region as the North American type of the sun-snaring myth (Luomala 1940: 20, 22), seem to strengthen this hypothesis, especially as most of the North American variants stress the darkness caused by the snaring of the sun. For three reasons, however, I doubt that the interpretation of the snare as the diminishing diurnal arc is the solely valid one: 1) The motive of snaring the sun and of resulting darkness is also known from regions far outside the polar circle, where the phenomenon of the polar night does not exist. 2) The sun-snaring myths of North America and of Oceania might give us some hints that it is the moon that is involved in the process of snaring the sun: in several North American versions, it is the rabbit—a well-known moon attribute—which snares the sun. In some Polynesian versions, the woman, of whose hair the snare was made, is identified with the moon. Therefore, it should also be taken into account that the process of snaring the sun might mean a solar eclipse. 3) The possibility should also be considered that the sun snaring might in some cases mean a simple sunset: On the Gilbert Islands, there exists a string figure called 'the sun' that was performed to help the sun set (Grimble 1931: 216).

The suggested interpretations for the sun-snaring motive can be summarized as follows: 1) solstice, especially winter solstice, 2) solar eclipse, 3) sunset. Whichever of these astronomical events is meant, it is the purpose of the sun snarer to manipulate the normal course of the sun: He wishes to prevent the disappearance of the sun for the polar night; he causes a solar eclipse; or he tries to fasten or to slow down the process of sunset. In each case the sun snaring means a manipulation of the cosmic order. Furthermore, we can differentiate between the manipulation of the spatial cosmic order (see 1.2.1) and of the temporal one (see 1.2.2, examples from Polynesia and the taboo of arctic people against playing string figures).

Two different functions of ropes concerning the cosmic order could be shown: 1) ropes establishing the cosmic order and 2) ropes used to manipulate the cosmic order. According to our rational thinking, such connections between ropes and the cosmic order are not self-evident, but let us have a look at:

2. Ropes and threads as magical instruments

Several languages have kept the inner connection between ropes and magic reflected in the etymology of the words for 'string' as well as for 'enchantment' (Scheftelowitz 1912: 17). In magical and symbolical thinking three aspects of ropes are stressed most: 1) They provide the means to ascend and thus enable humans to reach transcendental regions. In mythic times the ascent had been possible for all men, but now only men with high spiritual power succeed in finding the rope for the ascent. 2) Ropes link things with each other and with the first cause of existence and enable the exchange of power and life essence. Men who have knowledge of these ropes are the initiated ones of their culture (Eliade 1961: 119). 3) Ropes have the function of binding and loosening. The very conflicting nature of this function is reflected in many myths and magical rites (Scheftelowitz 1912: 3-12, 27-38; Eliade 1986: 116-122). The act of binding can mean punishment, enslavement, illness, or even death for the bounded one, but it can also tie him to the transcendental ground of his existence and thus integrate him in the cosmic order. Analogously, the act of loosening can mean physical and spiritual liberty, health, and life to the loosened one, but also loss of his transcendental ground and the unleashing of chaos. As the act of binding and loosening has such important meaning for the bodily and spiritual well-being of all living beings and the regulation of the cosmic order, the power of binding and loosening is a quality only of the highest gods (e.g. the Indian god Varuna) or god-like heroes and kings. In an Indian myth the ability of tying the sun proves a man to be the highest god (Rivers 1967: 206). In the Old Testament, God illustrates the deep gap between his own power and that of Job by asking him (Job 38, 31): "Do you bind the ties of the Pleiades, or do you loosen the ties of Orion?" Another example from the Bible shows us the power of binding and loosening as a means for establishing moral order. When Christ appointed Peter as the head of the church, he told him: "Whatever you shall bind on earth shall be bound in heaven." (Matt. 16:19). In the North American sun-snaring myths a basic motivation was to punish the sun for its bad behaviour. This means that binding and loosening something, whether human beings or celestial bodies, has also the figurative meaning of establishing moral or spiritual order.

Summary

Myths from different parts of the world give examples of ropes that build up the different elements of the coordinate grid of space and time. They also show examples of ropes being used to manipulate the established order. The strange idea that ropes and snares might be responsible for the polar night, solar eclipses, and other manipulations of the celestial bodies' regular movements becomes clearer in the context of examples of the magic use of ropes: In magical thinking ropes establish the cosmic order in a much more extended sense than just the astronomical one. Cosmic order means astronomical order as well as social and ethical order. By the use of ropes initiated persons like gods, shamans, heros, or kings are able to manipulate the cosmic order in its different aspects. The process of binding and loosening means transformation of conflict into law and order, of chaos into cosmos.

*Bahnhofstrasse 1, 82205 Gilching-Geisenbrunn, Germany, mr@infis.org.

References

Alexander, H. B. 1964. *Latin-American mythology* (Mythology of All Races 11). New York.
Bronsart, H. von 1963. *Kleine Lebensbeschreibung der Sternbilder.* Stuttgart.
Dechend, H. von and de Santillana, G. 1993. *Die Mühle des Hamlet. Ein Essay über Mythos und das Gerüst der Zeit.* Berlin.
Diederichs, U. 1998. *Die schönsten Märchen vom Mond.* Munich.
Dixon, R. B. 1964. *Oceanic mythology* (Mythology of All Races 11). New York.
Eliade, M. 1961. Mythes et symboles de la corde. *Eranos-Jahrbuch* **29**, 109-137.
Eliade, M. 1986. *Ewige Bilder und Sinnbilde: über die magisch-religiöse Symbolik.* Frankfurt am Main.
Griaule, M. 1980. *Schwarze Genesis. Ein afrikanischer Schöpfungsbericht.* Frankfurt am Main.
Grimble A. 1931. Gilbertese astronomy and astronomical observances. *Journal of the Polynesian Society* **40**, 197-224.
Häny, A. (ed.) 1987. *Die Edda. Götter- und Heldenlieder der Germanen*, 2nd ed. Zürich.
Harva, U. 1933. *Religiöse Vorstellungen der altaischen Völker.* Porvoo/Helsinki.
Hatt, G. 1949. *Asiatic influences in American folklore.* Copenhagen.
Hirsch, F. 1965. *Der Sonnwendbogen.* Göttingen.
Hudson, T. and Underhay, E. 1978. *Crystals in the sky: An intellectual Odyssey involving Chumash astronomy, cosmology and rock art.*
Jacobi, L. 1981. *Vom Werden der Welt und des Menschen: Schöpfungs- und Entstehungsmythen der Völker.* Schaffhausen.
Jakubassa, E. 1998. *Märchen aus Neuseeland.* Augsburg.
Krupp, E. C. 1983. *Echoes of the ancient skies: The astronomy of lost civilizations.* New York, Oxford.
Luomala, K. 1940. *Oceanic, American indian, and African myths of snaring the sun* (Bernice P. Bishop Museum Bulletin 168). Honolulu.
MacDonald, J. 1998. *The arctics sky: Inuit astronomy, star lore and legend.* Toronto.
Milbrath, S. 1999. *Star gods of the Maya: Astronomy in art, folklore, and calendars.* Austin.
Shorey, P (tr.). 1987. *Plato. The Republic*, vol. 2 (Loeb Classical Library 276). Cambridge MA and London.
Rivers, W. H. R. 1967. *The Todas.* London.
Scheftelowitz, I. 1912. *Das Schlingen- und Netzmotiv im Glauben und Brauch der Völker.* Gießen.
Wilson-Haffenden, J. R. 1927-1928. Ethnological notes on the Kwottos of tot (Panda) district, part III. *Journal of the African Society* **27**, 142-154.

L'astrologie et le comput dans les romans de Chrétien de Troyes: *Erec et Enide, Le Chevalier de la Charrette*

Rafal Perkowski*

Abstract
Chrétien de Troyes, auteur du XIIe siècle, écrit ses romans pour les cours de Champagne et de Troyes: centres aristocratiques et culturels très importants. Le présent article propose une recherche des origines de deux images utilisées par Chrétien: le premier d'*Erec et Enide* et le second du *Chevalier de la Charrette*. Nous essayons de concevoir ces images dans l'ensemble des 'habitudes intellectuelles' et par les biais de l'*Imago Mundi* du XIIe siècle. Les textes des romans de Chrétien sont comparés au comput de Philippe de Thaon et au traité astrologique d'Albumasar. Le mode de notre argumentation est issu de la méthode d'histoire des idées (Lovejoy 1933). Nous supposons que les notions du comput, et le savoir astrologique, peuvent expliquer certains aspects de la structure des romains et du choix des images littéraires.

Dans cet article je tenterai d'exposer les raisons pour lesquelles Chrétien de Troyes, auteur du XIIe siècle, introduit deux figures singulières dans ses romans. La première image est employée par Chrétien au commencement d'*Erec et Enide*: le jour de Pâques le roi Arthur veut chasser le cerf blanc pour restaurer la coutume (Fritz 1994b: *Erec et Enide* 37–38). La seconde se trouve dans le roman *Le Chevalier de la Charrette*: au temps de l'Ascension, Lancelot accepte de monter sur la charrette des condamnés afin de retrouver la reine enlevée par Méléagant (Fritz 1994a: *Chevalier de la Charrette* 30-378). Je voudrais expliquer l'apparition du cerf et du chevalier de la charrette en les replaçant dans l'ensemble des idées constituant l'image du monde—*Imago mundi* au XIIe siècle, en utilisant la pensée et les textes de l'environnement intellectuel de l'auteur. Je mènerai mon examen conformément à la méthode de l'histoire des idées, donc j'essayerai de comprendre le texte du point de vue des 'habitudes intellectuelles' typiques des contemporains de Chrétien. Je dois faire remarquer que mon parcours est très sélectif et que je le considère comme une proposition méthodologique plutôt qu'une solution définitive au problème.

Chrétien, restant au service des cours champenoise et troyenne, vit et construit ses romans dans les milieux princiers, au cœur de la diffusion des idées savantes, tant celles héritées du haut moyen âge que celles apparues avec les traductions arabo-latines au cours du XIe et XIIe siècles. Les seigneurs de Champagne et de Troyes subissaient également l'influence du phénomène que l'on a habitude d'appeler la renaissance du XIIe siècle (Benton 1961: 555-557, 573-575).

Si l'on enquête sur les cours princières, on note d'après Pierre Riché que les princes avaient toujours de l'intérêt pour les études astrologiques, et voulaient saisir le Cosmos (Riché 1989: 271), et cette attitude propre au Moyen Age de Charlemagne et de ses descendants est toujours vive au XIIe siècle. Patricia Stirnemann (1989: 177) remarque un progrès de l'alphabétisation de la haute aristocratie au XIIe et XIIIe siècles, qui d'un côté favorise la production littéraire en français, de l'autre la traduction des ouvrages en latin, y compris 'scientifiques'. Parmi ces derniers se trouvent de traités portant sur le Cosmos.

Les Pâques et l'Ascension évoqués par Chrétien sont les solennités du temporial et leurs dates peuvent être déterminées grâce au comput. Ainsi on observe que Chrétien situe dans le temps ses histoires en s'appuyant sur le système du calcul médiéval servant à trouver la date des Pâques. Je n'aborderai pas l'histoire du calcul et je n'évoquerai pas non plus les détails techniques, sauf les aspects indispensables pour mon propos. Cette méthode de calcul a été établie par Bède le Vénérable; pour préciser, je noterai que la notion 'computus' désignait tant la méthode du calcul que le texte la décrivant (Borst 1993: 29). Le nombre de manuscrits consacrés au comput est abondant, l'apprentissage de ce raisonnement est introduit dans le *corpus* des arts libéraux et le niveau de sa maîtrise paraît très élevé (Riché 1989: 271). C'est sans aucun doute que les clercs étaient familiarisés avec la matière de ce calcul particulier. Chrétien était probablement clerc ou au moins il connaissait les arts du *quadrivium*. En décrivant l'étoffe d'Erec il fait un louange de l'astronomie (Fritz 1994b: d'*Erec et Enide* 6730-6782). Le comput est aussi présent dans les livres de prières des laïcs, et l'on en trouve un bel exemple dans le psautier du XIIe siècle de Christina de Markyate (L'Hermite-Leclercq 1991: 31).

Mais il semble que le public, qui ne pouvait pas l'étudier en latin, prenait connaissance de ce calcul en participant à la liturgie chrétienne plutôt qu'en lisant les ouvrages en langue vernaculaire, comme le premier comput en langue française de Philippe de Thaon, ainsi que le suppose Mary Dominica Legge (1963: 21-22). Philippe traduit le texte en français pour qu'il soit utilisé dans les sermons et affirme l'utilité de son entreprise pour les clercs mais aussi pour les laïcs (Short 1984: 7 [de Philippe de Thaon 224-225]). Il faut cependant constater que l'apprentissage du comput s'effectuait toujours en latin.

Néanmoins l'étude du comput composé par Philippe de Thaon au XIIe siècle enseigne que malgré son apparence très arithmétique (Flamant 1984: 31-43), il n'échappe pas à l'instruction théologique, philosophique et cosmologique. Philippe, en s'appuyant sur l'autorité de Saint Augustin, enseigne qu'il faut prêter l'attention aux fêtes pour maintenir les lois (Short 1984: 5 [Philippe de Thaon 33-38]). Saint Augustin, *auctoritas* indiscutable des computistes, rejetant l'astrologie prophétique, accorde un rôle important à la contemplation du ciel pour admirer l'œuvre de Dieu émanant dans la réalité temporelle (Collish 1997: 30) et pour célébrer les fêtes (McCluskey 1998: 32). Philippe ne se borne pas à fournir les informations astronomiques pour donner l'étymologie des noms des jours et des mois, mais il fait un point indépendant sur l'image du Cosmos: les planètes, les signes du zodiaque et leur signification allégorique. On peut apercevoir dans cet exposé les traces du rôle considérable joué par le comput dans les sciences physiques et mathématiques jusqu'à l'apparition des traductions arabes (Lejbowicz 1992: 194). Les problèmes de mesure du temps impliquaient les questions astronomiques et astrologiques. Philippe souligne aussi les liens des computistes et des astronomes (Short 1984: 8 [Philippe de Thaon 305-306]); donc le temps et sa mesure sont liés aux révolutions des corps célestes dans le texte fondamental pour la formation des esprits médiévaux. Par ce lien profond et explicite, les grands luminaires, les planètes, les astres et les constellations sont à la fois les gardiens et les indicateurs du temps. Bref, le comput de Philippe de Thaon dévoile l'association de deux idées: celle du temps et celle du ciel. Même en dehors du comput, cette relation semble s'étaler sur plusieurs domaines de l'activité intellectuelle et constituer une des 'constantes mentales' du XIIe siècle et du Moyen Age plus généralement (Lecoq 1992: 113-124).

Je me demande donc si cette 'habitude mentale' se trouve chez Chrétien. Chrétien nous indique les moments, les fêtes liturgiques, mais est-ce que les points du temps ont leurs homologues dans l'espace céleste? Pour saisir cela, on doit parcourir les textes classiques étudiés par les clercs, surtout les textes d'Ovide: auteur des œuvres portant sur la mythologie, l'astronomie et le calendrier. Grâce à l'abondance de manuscrits et l'influence manifeste sur les esprits médiévaux, certains chercheurs accordent à la période examinée le nom de 'l'âge d'Ovide' (Schilling 1992: LII). Comme nous allons le voir, l'étude de la structure des *Fastes* d'Ovide apporte une leçon instructive. Les *Fastes* constituent une sorte de calendrier où les histoires mythologiques et les coutumes sont disposées selon les catégories du temps du calendrier julien, mais aussi et surtout selon l'ordre astronomique (Gee 2000: 1). Cette attitude se radicalise même dans le traité rhétorique de Quintilien (Cousin 1975: 79 [*Institution oratoire* 1.4.4]) où la poésie ne peut pas être comprise sans l'astronomie qui sert à indiquer le temps des histoires (Gee 2000: 21). Ovide est évoqué par Philippe de Thaon. Pour Chrétien, Ovide est le maître. Il l'imite dans son oeuvre *Philomène*, il le mentionne aussi dans le prologue de *Cligès.* Même si Chrétien ne cite pas les *Fastes*, il peut les imiter, dans sa structure, en procédant à la 'technique allusive', connue dans l'Antiquité et au Moyen Age. Cette technique rhétorique pouvait servir aux auteurs en langues vernaculaires à regagner l'autorité et à rendre hommage à leur modèle (Mora 2001: 219- 224). L'on arrive de cette manière à la constatation: parmi les modèles littéraires de Chrétien, il s'en trouve un qui lie le savoir astronomique, la mesure du temps et les histoires.

Je présente une image qui m'a amené à comparer le texte du roman avec les travaux d'Albumasar, astrologue arabe du IXe siècle (Sniezynska-Stolot 1997). Il me s'agit de la scène mentionnée au début avec Lancelot et la charrette. La trace qui m'a permis de m'intéresser aux liens probables des romans avec l'astrologie est le nom du héros, Lancelot, et le nom de la constellation. A l'horizon céleste apparaît la constellation qui s'appelle en latin *Lanceator. Lanceator* est présenté dans l'iconographie astrologique comme homme assis dans le chariot. Cette figure est très fréquente dans l'iconographie astrologique. On rencontre ce type de représentation dans les enluminures d'*Aratea* (Le Boeuffle 1975). Puis, l'image de l'homme dans le chariot se trouve parmi les illustrations du manuscrit d'Abumasar, dans sa version abrégée de Fendulus. Dans la première moitié du XIIe siècle, le remanieur de la traduction latine d'Albumasar faite par Hermann de Carinthie, Fendulus (Georgius Zathori Zapari Fenduli), a enrichi le texte de miniatures. (Clark 1979: 32-34). De même la figure du cerf représente la constellation qui porte le même nom. Donc les types iconographiques présentés dans les manuscrits astrologiques ressemblent beaucoup aux topoi utilisés par Chrétien. Les images correspondant aux scènes mentionnées se trouvent dans les manuscrits d'Albumasar (groupe de Fendulus) et figurent parmi les représentations des *paranatellonta*, cela veut dire, des constellations qui accompagnent les décans des signes du zodiaque. L'image du Cerf (la constellation *Cervus*) est associée avec le premier et deuxième décan du signe du Bélier, puis avec le troisième décan du Scorpion (Sniezynska–Stolot 1997: figs. 16-19, 63-64). *Lanceator*, comme homme sur la charrette, accompagne le troisième décan du Bélier, le premier décan des Gémeaux, et le troisième décan du Lion (Sniezynska-Stolot 1997: figs. 26-29, 44-45).

Néanmoins de cette similarité peut naître des doutes. Les recherches sur l'histoire de l'astrologie étant loin d'être achevées, obligent à la prudence dans les conclusions. Sniezynska-Stolot remarque une situation pareille dans la recherche concernant l'iconographie astrologique (Sniezynska-Stolot 1994, 9). Alors mes conclusions qui ne prétendent pas être définitives représentent un état du savoir accessible à un moment donné du progrès des deux disciplines. La difficulté résulte également du fait que les manuscrits qui servent de soutien pour mon hypothèse étaient réalisés à partir du XIIIe siècle, tandis que le texte analysé date de la fin du XIIe siècle. Il reste cependant une hésitation qui porte sur la source des enluminures des manuscrits du groupe de Fendulus. Les chercheurs ne savent pas encore si les témoins des manuscrits sont les premiers exemples de l'illustration du traité d'Albumasar, s'il existait un modèle fait après 1140, dans le sud de l'Italie, d'après lequel on a confectionné les manuscrits sauvegardés (Sniezynska-Stolot 1994, 9). Le problème du rapport entre le texte et l'image révèle des points instructifs. L'arrangement des images dans le traité de Fendulus dépend parfois des goûts esthétiques plutôt que de la fidélité au texte du traité (Clark 1979: 112). Pourtant, la comparaison avec un autre groupe de manuscrits astrologiques, celui d'Abraham Ibn Ezra, auteur qui ne connaissait pas probablement la version de Fendulus, et qui est plus fidèle à l'égard du texte (Sniezynska-Stolot 1998: 16), apporte quelques changements, mais pour la majorité ressemble à la classification du groupe de Fendulus (Table I).

Constellation	Le groupe de Fendulus	Le manuscrit d' Abraham ibn Ezra
Cervus	premier et deuxième décans du Bélier troisième décan du Scorpion	deuxième décan du Bélier
Lanceator	troisième décan du Bélier premier décan des Gémeaux troisième décan du Lion	troisième décan du Bélier premier décan des Gémeaux troisième décan du Lion

Table 1. La comparaison entre les représentations astrologiques dans les manuscrits de Fendulus et les traités d'Abraham Ibn Ezra.

Alors, dans certains cas on peut constater que la tradition iconographique de l'emplacement des figures dans le cercle zodiacal est bien fixée. Surtout dans le cas de *Lanceator* qui dans les deux groupes de manuscrits occupe la même place et sa façon de sa présentation est pareille.

J'ai essayé d'établir les similarités au niveau des apparences: les images qu'on aperçoit chez Chrétien et leurs homologues iconographiques dans les traités astrologiques. A dessein de comprendre la logique de l'intercalation de ces images et leur provenance astrologique possible il faut s'arrêter sur le comput qui est lié avec l'astronomie et l'astrologie.

L'histoire d'*Erec et Enide*, commence le jour de Pâques avec la chasse au cerf (Fritz 1994b: *Erec et Enide* 27-38). Puis les événements déroulent autour de Pentecôte: les noces d'Erec à Caradigan (Fritz 1994b: *Erec et Enide* 1916-1924) et le tournoi sous Danebroc (Fritz 1994b: *Erec et Enide* 231-233). Le roman s'achève avec le couronnement d'Erec et d'Enide à Nantes à la Nativité (Fritz 1994b: *Erec et Enide* 6551-9650). Dans d'autres termes, elle débute au moment de la résurrection de Christ et finit le jour où il faut célébrer sa naissance.

La partie majeure du *Chevalier à la Charrette*, 5270 vers des 7112 (Fritz 1994a), s'effectue entre l'Ascension et la Pentecôte. Chrétien amplifie cette période de dix jours aventureux ce qui correspond à dix jours qui décalent la Pentecôte de l'Ascension selon la liturgie chrétienne.

Dans le contexte du calendrier perpétuel médiéval, je me réfère au calendrier contemporain à la *Massa Compoti* d'Alexandre de Villedieu (Van Wijk 1936: 144-155) établi 30 ans après l'apparition d'*Erec*, les fêtes dont Chrétien fait l'usage dans les deux romans, se présentent de la manière suivante:

18 mars	Soleil dans le Bélier
21 mars	Equinoxe du Printemps; *Primus terminus* de Pâques
17 avril	Soleil dans le Taureau
18 avril	*Ultimus terminus* de Pâques
30 avril	*Primus dies* de l'Ascension
9 mai	*Primus terminus* de la Pentecôte
18 mai	Soleil dans les Gémeaux
3 juin	*Ultimus dies* de l'Ascension
6 juin	*Ultimus terminus* de la Pentecôte

Table 2. Les clés du comput (le premier et le dernier jour où on peut célébrer les fêtes) et les jours où le soleil entre dans les signes zodiacaux convenables, d'après le calendrier perpétuel du temps d'Alexandre de Villedieu (Van Wijk 1936: 143-145).

Les textes du comput enseignent que le dimanche de Pâques suit le quatorzième jour de la lune pascale et ne peut pas précéder l'équinoxe du printemps. L'équinoxe a lieu au mois de mars où *la surrectiun/De Jesu Crist, le reis/Primes fut en cest meis* (Mall 1873: Philippe de Thaon 860-862). L'Ascension vient quarante jours après la commémoration de la mort et de la résurrection du Christ. Mais la date de l'équinoxe ne semble pas être fixée par les computistes mêmes. Se développe un grand débat qui durera jusqu'à la réception par l'occident chrétien de la science arabe et de son instrument célèbre, l'astrolabe. Dans les descriptions de cet outil astrologique dont l'usage devient très fréquent au cours du XIIe siècle, on voit le moment de l'équinoxe du printemps ajusté à l'entrée du soleil dans le signe astronomique du Bélier (McCluskey 1998: 178-180). En conséquence, on peut constater que la chasse au cerf se rapporte d'une certaine manière à l'équinoxe du printemps ou au moins dépend directement de ce phénomène astronomique, et celui-ci a lieu au moment où le soleil commence son parcours annuel au signe du Bélier. L'histoire du Lancelot doit venir quarante jours après.

J'ai comparé les donnés astrologiques et computistiques qui m'ont conduit à établir le schéma général suivant:

Le décan de signe du zodiaque	La constellation accompagnant le signe dans le décan (*paranatellonta*)		Les solennités du comput		
Premier décan de Bélier	Cervus		Pâques		
Deuxième décan de Bélier	Cervus*				
Troisième décan de Bélier					
Premier décan de Taureau					
Deuxième décan de Taureau				Ascen-sion	
Troisième décan de Taureau		Lanceator			
Premier décan de Gémeaux		Lanceator			Pente-côte
Deuxième décan de Gémeaux					
Troisième décan de Gémeaux					

Table 3. Les relations entre l'apparition des constellations à l'horizon céleste selon Fendulus, et les termes du comput selon Alexandre de Villedieu. *Uniquement dans le groupe de Fendulus.

D'après le schéma proposé, on voit que les constellations s'associent, vaguement, avec les fêtes qui structurent les romans de Chrétien. Les images des constellations pouvaient servir de signes du temps. Les formes fabuleuses des constellations qui conditionnaient les calculs et la mesure du temps, indiquent les points dans l'espace céleste mais aussi les moments dans le temps.

Le roi Arthur veut chasser le cerf le jour de Pâques, donc après l'équinoxe de printemps. Dans les manuscrits d'Albumasar la constellation du cerf accompagne le premier et le deuxième décan de Bélier, donc les décans qui se trouvent proche de l'équinoxe de printemps si l'on suit le parcours annuel du soleil, planète importante pour le comput (Sniezynska-Stolot 1997). L'Ascension est établie quarante jours après Pâques, et dans cette période Lancelot voyage dans un chariot. Si l'on examine les manuscrits d'Albumasar, on voit que le personnage assis dans un chariot est connecté avec le troisième décan de Taureau. De la sorte, similairement à la figure du cerf, l'homme dans le chariot, *Lanceator,* correspond à la période où l'on peut célébrer l'Ascension. Il faut cependant noter que Chrétien ne fait aucune allusion explicite à l'équinoxe, et Philippe de Thaon, de son côté, n'introduit ni constellation du *Cervus* ni celle du *Lanceator* et de la Couronne dans le texte du comput. A mon avis, c'est le traité d'Albumasar qui peut prouver un tel lien .

Mon hypothèse montre la possibilité d'utiliser deux sources d'invention à la fois. Les deux documents médiévaux, le comput et les traités d'astrologie formaient les esprits de l'époque. En conséquence ils pouvaient nourrir l'imagination du poète qui se trouvait à la recherche du temps afin de donner l'ordre pour les histoires. Surtout dans le temps où le calendrier uniforme n'existe pas encore.

Il faut cependant souligner que Chrétien est poète, donc même si on peut comprendre certains images dans leur entourage 'scientifique', Chrétien est poète, il utilise les topoi qui existent déjà dans la littérature. Ceci donne une image intellectuelle de l'époque où vit Chrétien et indique deux habitudes d'étudier ou de décrire le ciel, l'une computistique trouvant ses racines dans le moyen âge de Bède le Vénérable, l'autre 'nouvelle' avec le parfum arabe de nouvelles traductions des oeuvres astrologiques. La structure du roman peut venir de la structure du comput, alors que le réservoir des images paraît dériver du traité d'Albumasar (Sniezynska-Stolot 1997). Puis Chrétien suit probablement son maître Ovide et essaye d'introduire l'ordre de la sphère céleste dans le temps de la narration.

En guise de résumé, je voudrais faire deux remarques. Premièrement, parmi les manuscrits des romans interrogés on repère des enluminures du cerf et de l'homme de la charrette. Cela veut dire que les scènes pouvaient être importantes du point de vue de la mnémotechnique. Elles pouvaient exercer la fonction d'*imagines agentes* de la quatrième partie de l'art de rhétorique, des images mettant en action la mémoire. Je rappelle le fragment, souvent repris par des savants médiévaux, de définition de l'image d'Isidore de Séville: *dès que l'on voit l'image, on se souvient de la chose* (De Bruyne 1998: 92) Les travaux de Frances Amelia Yates et ceux plus récents de Mary Carruthers privilégient le système mnémotechnique dans la transmission du savoir jusqu'à l'âge de l'imprimerie. Puis on connaît le système de Métrodore qui conseillait de mémoriser en utilisant le cercle zodiacal, avec ses décans (Yates 1966: 36). On voit aussi que le comput de Philippe de Thaon, précédé par le bestiaire, révèle la fonction mnémotechnique du texte. Mary Carruthers s'interroge sur le rôle du cercle zodiacal dans l'apprentissage du comput et du bestiaire (Carruthers 1990: 126). Ainsi, l'emploi des figures dans les romans de Chrétien pourrait venir de cette tradition.

La deuxième remarque est issue du travail d'Arno Borst. Au moment de développement des langues vernaculaires, il s'est produit une confusion entre le latin *computare* et le français 'conter' (Borst 1993: 46). Donc, pour certains, conter pouvait signifier calculer et chercher le temps.

*Université Jagellonne de Cracovie, Université de Paris IV, Paris–Sorbonne.

Bibliographie

Benton, J. F. 1961. The court of Champagne as a literary centre. *Speculum* **36**, 551-591.

Borst, A. 1993. *The ordering of time: From the ancient Computus to the modern computer.* Cambridge, Oxford.

Carruthers, M. 1990. *The book of memory: A study of memory in Medieval culture*. Cambridge.

Clark, V. A. 1979. *The illustrated 'Abridged Astrological Treatises of Albumasar': Medieval astrological imagery in the west* (doctoral dissertation submitted at the University of Michigan). Ann Arbor.

Collish, M. 1997. *Medieval foundations of the western intellectual tradition 400-1400*. New Haven and London.

Cousin, J. (ed.) 1975. *Institution oratoire*, vol. 1. Paris.

De Bruyne, E. 1998. *Etudes d'esthétique médiévale*, vol. 1. Paris.

Flamant, J. 1984. Temps sacré et comput astronomique. In *Le temps chrétien de la fin de l'antiquité au Moyen Age. IIIe–XIIe siècles* (Colloques internationaux du Centre National de la Recherche Scientifique 604), 31-43. Paris.

Fritz, J.-M. (ed.) 1994a. *Chevalier de la Charrette*. In *Chrétien de Troyes. Romans*. Paris.

Fritz, J.-M. (ed.) 1994b. *Erec et Enide*. In *Chrétien de Troyes. Romans*. Paris.

Gee, E. 2000. *Ovid, Aratus, and Augustus: Astronomy in Ovid's Fasti*. Cambridge.

Le Boeuffle, A. (ed.). 1975. *Les phénomènes d'Aratos [attribué à] Germanicas*. Paris.

Lecoq, D. 1992. Le temps et l'intemporel sur quelques représentations médiévales du monde au XIIe et au XIIIe siècles. In *Le temps, sa mesure et sa perception au Moyen Age (Actes du Colloque, Orléans 12-13 avril 1991)*, ed. B. Ribémont, 113-149. Caen.

L'Hermite-Leclercq, P. 1991. Le psautier de Christina de Markyate. In *Prier au Moyen Age. Pratiques et expériences (V^e^-XVe siècles)*, ed. N. Bériou, J. Berlioz, and J. Longère, 31-34. Paris.

Legge, M. D. 1963. *Anglo-Norman literature and its background*. Oxford.

Lejbowicz, M. 1992. Computus. Le nombre et le temps altimédiévaux. In *Le temps, sa mesure et sa perception au Moyen Age (Actes du colloque, Orléans 12-13 avril 1991)*, ed. B. Ribémont, 151-195. Caen.

Lovejoy, A. O. 1933 and later printings. *The great chain of being: A study of an idea*, Cambridge.

Mall, E. 1873. *Li compoz Philipe de Thaün*. Strasbourg.

McCluskey, S. C. 1998. *Astronomies and cultures in Early Medieval Europe*. Cambridge.

Mora, F. 2001. Remplois et sens du jeu dans quelques textes médio-latins et français des XIIe et XIIIe siècles: Baudri de Bourgueil, Hue de Rotelande, Renaut de Beaujeu. In *Auctor & Auctoritas: Invention et conformisme dan l'écriture médiévale. Actes du colloque de Saint-Quentin-en-Yvelines, 14-16 juin 1999* (Mémoires et Documents de l'École des Chartes 59), ed. M. Zimmermann, 219-230. Paris.

Riché, P. 1989. *Ecoles et enseignement dans le Haut Moyen Age*. Paris.

Schilling, R. (ed.) 1992. *Les Fastes. Ovide*, vol. 1. Paris.

Short, I. (ed.) 1984. *Comput* (MS BL Cotton Nero A. V). *Phillippe de Thaon*. London.

Stirnemann, P. 1989. Les bibliothèques princières et privées aux XIIe et XIIIe siècles. In *Histoire des bibliothèques françaises. Les bibliothèques médiévales. Du VIe siècle à 1530*, ed. A. Vernet, 173-191. Paris.

Sniezynska-Stolot, E. 1994. *Ikonografia znakow zodiaku i gwiazdozbiorow w sredniowieczu (Iconographie des signes du zodiaque et des constellation au Moyen Age)*. Krakow.

Sniezynska-Stolot, E. 1997. *Ikonografia znakow zodiaku i gwiazdozbiorow w rekopisie Albumasara (Iconographie des signes du zodiaque et des constellation dans le manuscrit d'Albumasar)*. Krakow.

Sniezynska-Stolot, E. 1998. *Ikonografia znakow zodiaku i gwiazdozbiorow w rekopisie monachijskim Abrahama ibn Ezry (Iconographie des signes du zodiaque et des constellation dans le manuscrit d'Abraham Ibn Ezra)*. Krakow.

van Wijk, W. E. (tr.) 1936. *Le nombre d'or. Etude de chronologie technique suivi du texte de la Massa Compoti d'Alexandre de Villedieu, avec traduction et commentaire*. La Haye.

Yates, F.A. 1966. *The art of memory*. London.

English summary: Astrology and computus in the Chrétien de Troyes novels

As the topic of my article I have chosen the role of the astrological treaties and the computus in the introduction of cosmological ideas to literature, especially to French literature before the end of the 12th century.

I have selected two images from the Chrétien de Troyes novels, *Erec et Enide* and *Chevalier de la Charrette*. The first picture represents a stag hunt at Easter time arranged by King Arthur (Fritz 1994a: *Chevalier de la Charrette* 37-38), and the second one, Lancelot sitting in the cart on Ascension Day (Fritz 1994b: *Erec et Enide* 30-378).

I would like to emphasise that I will present a methodological proposition rather than a definitive solution to the problem. I use the methodology of the History of Ideas (Lovejoy 1933), so I try to understand the text in the context of 'mental habits' specific for the 12th-century world view.

Easter and the Day of the Ascension, mentioned by Chrétien, are also in the computus, the medieval system used to determine Easter Sunday. In the computus literature we find the information that Easter is connected to the vernal equinox (Easter can not be celebrated before the equinox and the 14th day of the Jewish lunar month *Nisan*), and Ascension Day comes forty days after Easter Sunday. The equinox seems to be one of the key points in this method of time reckoning. In the description of Arabic astrolabes (the instrument used for measuring the position of celestial bodies, employed by 'computus specialists' in the 12th century), one can distinguish the moment of the vernal equinox adjusted to the entrance of the sun into the astronomical sign of Aries. The aim of the computus required astrological knowledge. Scholars of liberal arts studies had to learn the computus at the beginning of their education. I suppose that Chrétien acquired the knowledge of the *quadrivium* (Fritz 1994b: *Erec et Enide* 6730-6782). He also acquired knowledge of the method of computing time according to the computus. According to the opinion of researchers who have been involved in the study of the computus, this science was joined with astrology. This affirms that those disciplines were very close in the medieval mentality.

We may thus conclude that Chrétien, who conjures up the notions of the computus, probably used astronomical data to get the narration in the same order as Ovid in the *Fasti*. The "Fasti of Ovide is syncrasis of the Roman religious calendar, Julian panegyric, and astronomical material, the rising and settings of stars in relation to the Roman year" (Gee 2000: 1). Ovid was an author very much appreciated by Chrétien de Troyes, who imitates him, especially in the *Philomène*. The 12th century is also called *Aetas Ovidiana* due to the large influence of this author.

However, to my mind, the astronomical data for Chrétien's 'time reckoning' can be located in the astronomical treaty of Albumasar (Sniezynska-Stolot 1997). Albumasar was a very famous Arabic astrologer of the 9th century, but the first Latin translations of his work came out during the 12th century. He had written the synthetic astrological compilation based on the Greek, Persian, and Indian astrological traditions. He had presented the twelve signs of the zodiac and its thirty-six decans in textual and pictorial form. He had mentioned not only the zodiacal constellations but also other non-zodiacal constellations (paranatellonta).

In his writings we find the images of the stag (constellation of *Cervus*) and of the man sitting in the cart (constellation of *Lanceator*). They are connected with the signs of Aries and Taurus. Therefore we can deduce that the astronomical distance between the appearance of the constellation of *Cervus* and *Lanceator* is equal to forty solar days. According to the computus handbooks the same interval lies between Easter Sunday and Ascension Day. But I underline the fact that the notion of the equinox is missing in the Chétien novels, and there are no remarks about the stag and the man in the cart in Philippe de Thaon's *Computus*. (Short 1984) The connection of the literary topos and the astronomical data is possible thanks to treaties of Albumasar (Sniezynska-Stolot 1997). I sum up my observations in the following table.

Chrétien's novel	Image	Computus	Constellation	Astrological explanations
Erec et Enide	White stag	Easter Sunday	Cervus	1st and 2nd decan of Aries
Chevalier de la Charrette	Knight in the cart	Ascension Day	Lanceator	3rd decan of Taurus

In my opinion Chrétien took advantage of the cosmological knowledge present in the 12th century. He uses the computus in the structure of his novels, but for images he refers probably to Albumasar's works. However Chrétien mainly shows us the manner of describing and understanding the Cosmos in the 12th century.

On the other hand, I think that to describe and calculate time was possible thanks to the computus. The computus was based on astronomy. Astrology and astronomy are two faces of the same science. In Chrétien's century we can assume that time connected with astronomy/astrology existed not only as a necessity of practical life, but also as a mental habit or idea, which was explicitly expressed in literature.

Betrayed lovers of Ištar: A possible trace of the 8-Year Venus cycle in *Gilgameš* VI:i–iii

Arkadiusz Sołtysiak*

Abstract
This paper discusses the beginning of the VIth tablet of the Babylonian story about Gilgameš, which contains the list of lovers of the goddess Ištar. The possibility of their relationship to the constellations of the heliacal settings of Venus in the 8–year cycle is hypothesised.

The VIth tablet of the Babylonian story about Gilgameš is one of a few mythological sources in which astral motifs are explicitly expressed. It describes the battle of the hero against the Bull of Heaven, the creature without doubt representing the constellation Taurus (Sołtysiak 1999b: 291; 2001). However the tablet begins with a dispute between Gilgameš and the goddess Ištar, the dispute which made Ištar angry and, in consequence, caused the above-mentioned combat. Gilgameš refused Ištar's courtship in an offensive way and enumerated her six previous lovers who always got into trouble on her behalf. Here is the proper passage of the text:

> (...) And Ištar the princess raised her eyes to the beauty of Gilgameš.
> "Come to me, Gilgameš, and be my lover! (...)
> I shall have a **chariot** of lapis lazuli and gold harnessed for you,
> With wheels of gold, and horns of *elmešu*-stone.
> You shall harness *umu*-demons as great mules!
> Enter into **our house** through the fragrance of pine! (...)"
> Gilgameš made his voice heard and spoke, he said to Ištar the princess,
> "(...) Which of your lovers [lasted] forever?
> Which of your masterful paramours went to heaven?
> Come, let me [describe?] your lovers to you! (...)
> For **Dumuzi** the lover of your youth
> You decreed that he should keep weeping year after year.
> You loved the colourful ***allallu*-bird**, but you hit him and broke his wing.
> He stays in the woods crying »My wing!«.
> You loved the **lion**, whose strength is complete,
> But you dug seven and seven pits for him.
> You loved the **horse**, so trustworthy in battle,
> But you decreed the whip, goad, and lash for him, (...)
> You decreed endless weeping for his mother Sililu.
> You loved the shepherd, herdsman, and chief shepherd
> Who was always heaping up the glowing ashes for you,
> And cooked ewe-lambs for you every day.
> But you hit him and turned him into a **wolf**,
> His own herd-boys hunt him down and his dogs tear at his haunches.
> You loved **Išullanu**, your father's gardener,
> Who was always bringing you baskets of dates.
> They brightened your table every day;
> You lifted your eyes to him and went to him,
> »My own Išullanu, let us enjoy your strength,
> So put out your hand and touch our vulva!«
> But Išullanu said to you, »Me? What do you want of me?
> Did my mother not bake for me, and did I not eat?
> What I eat (with you) would be loaves of dishonour and disgrace,
> Rushes would be my only covering against the cold«.

> You listened as he said this, and you hit him, turned him into a frog? (*dallalu*),
> Left him to stay amid the fruits of his labours.
> But the pole? goes up no more, [his] bucket goes down no more.
> And how about me? You will love me and then [treat me] just like them!"
> When Ištar heard this, Ištar was furious, and [went up] **to heaven**.
> Ištar went up and wept before her father Anu (...)[1]

At first sight this fragment contains nothing that could be interpreted in astral categories, except the last reference to Ištar ascending to heaven. However the character of the Bull of Heaven in the following part of the tablet and the character of Ištar as the goddess of the planet Venus suggest that some astral motifs can be concealed in the deeper layers of the text. The following discussion will be focused on an attempt to discover and explain these motifs. Of course hermeneutic procedures of this kind never provide sure results, but even the less probable hypotheses may be interesting for the researchers of astronomy in culture.

A very important aspect is the date and the context of the analysed story. The combat of Gilgameš and the Bull of Heaven had been included already in the Sumerian prototype of this Babylonian text. However the initial dispute between Gilgameš and Inanna (Sumerian counterpart of Ištar) looks different in the older source. The tablet is badly broken in this passage, but it is obvious that the lovers of the goddess are not mentioned. Inanna invites the hero to her temple and intends to perform a sacred marriage ritual with him. Gilgameš refuses because he wished "to catch? mountain bulls, to fill the cowpens", and this answer makes Inanna angry.

The ritual of sacred marriage is attested in the sources from the 21th to the18th centuries BC and belongs to the tradition of Uruk, the center of the cult of Inanna/Ištar and the city of Gilgameš. There are a number of hymns and stories concerning the courtship of the goddess and Dumuzi, the god of grain associated with the constellation Orion,[2] with whom the king was identified during the ritual. The following mythological events are described in a set of texts. The story about the descent of Inanna/Ištar to the Underworld concerns the journey of the goddess associated with the internal conjunction of Venus (Heimpel 1982: 9). After her return to the earth, the goddess was forced to choose a substitute who would be abducted to the Underworld. She chose her husband Dumuzi, who tried to escape from the hands of the demons. Eventually seized, he returned, owing to the self-sacrifice of his sister Geštinanna. All these events and their background are summarised in Table 1.

Since the analysed passage of the story about Gilgameš was composed at the same time when the kings of southern Mesopotamia celebrated the ritual of the sacred marriage of Ištar and Dumuzi (as stated before, it was not included in the Sumerian prototype), the name of the first lover of the goddess clearly relates to the above-mentioned cycle of sources. But what is the meaning of the five other characters?

The beginning of the 2nd millennium BC brought great changes in Mesopotamia. After the fall of the Sumerian third dynasty from Ur and the invasion of the Amorites, the interest in astronomy increased considerably. Regular observations of Venus are first attested in this period, in tablet 63 of the great astral omina series *Enuma Anu Enlil*, called also the tablet of Ammiṣaduqa, from the name of the Babylonian king some of whose years of ruling are covered by the observations (Reiner and Pingree 1975). This tablet contains the first known theory of the motion of Venus as well. It is very

[1] The translation is by Stephanie Dalley (1989: 77–80); the words in boldface specify those elements in the story which are discussed here as possibly being associated with heavenly phenomena. The Sumerian sources quoted are in the translation by Jeremy Black et al. (1998-2001), if not stated otherwise.

[2] In later sources Dumuzi was related to the constellation called the Hired Man (*Aries*), but there are good reasons for assuming that in the 2nd millennium his astral attribution was the Shepherd of Anu (Orion), and even in a few later astronomical texts such association was suggested (Livingstone 1986: 154; Sołtysiak 1999a: 43–44).

likely that the 8-year cycle of the visibility of Venus was discovered in that period; it may be also hypothesised that this discovery would be interpreted in a mythological way.

Month	Solar year	Mythological and ritual events
XII = February/March	grain ripening	the courtship of Dumuzi and Inanna
I = March/April	grain completely ripe	**the sacred marriage during the New Year festival**
beginning of April	Aldebaran ↓	the death of Gugalanna, 'great bull of heaven'
	Venus ↓ once every 8 years	descent of Inanna to the Underworld
middle of April	Venus ↑ once every 8 years	the return of Inanna
	beginning of the harvest	Inanna gives Dumuzi into the hands of the demons
end of April	Orion ↓ →Hydra	change of Dumuzi into a snake and his escape
II–III = April–June	harvest in progress	Dumuzi escapes to succeeding hiding places
beginning of June	Hydra ↓ end of harvest	Dumuzi seized by the demons
III/IV = end of June	Orion ↑ sowing	**festival of dead; Dumuzi returns from Underworld**
V = July/August	Hydra ↑ germination	*abum* festival

Table 1. Associations of the agricultural and stellar calendar with the mythological and ritual cycle of Inanna and Dumuzi in the beginning of 2nd millennium BC. Notations: ↓ – heliacal setting; ↑ – heliacal rising; → – visibility on the western horizon. The two lines in boldface refer to annual festivals which were explicitly related to the story about Inanna and Dumuzi.

The love and the death of Dumuzi in the ritual of the sacred marriage was associated first with the conjunction of Venus and Orion, next with the internal conjunction of Venus, and finally with the heliacal setting of Orion. However such a phenomenon occurs only once in the 8-year cycle of Venus. There are still five internal conjunctions of Venus in this cycle, each on the background of another constellation. Let us check whether the list of the remaining five lovers of Ištar might correspond to the constellations heliacally setting after the internal conjuction of Venus in the same cycle. Of course one must bear in mind that the whole cycle shifts somewhat more than two degrees in each 8-year period, and the scheme must be compatible with the historical period when the story was composed. Assuming that the conjunction with Orion is the point of reference, the other constellations should be distant from each other by 360° ÷ 5, or about 72° or its multiples.

The complete list of the Mesopotamian names of stars and constellations can be found in the series *mulApin*, a kind of astronomical compendium composed around the end of the 2nd millennium, but surely containing the elements of an older tradition. Short commentaries on some names are also provided there. In contemporary and later sources more particulars about the most important stars and constellations are present.

The *allallu*–bird is the second lover of Ištar. Its name never appears as the name of a star or constellation; the bird itself was identified as *Coracias garrulus* (Campbell Thompson 1924). The third lover is the lion, which can be identified as the constellation Leo. It fits the system well since Leo and Orion were about 72° from each other. Moreover, the lion is associated with Ištar as her animal attribute. In astronomical sources the name of Leo was written as Urgula or Urmah (both mean 'the great lion'). Next, the horse appears on the list of lovers. The Horse in the list *mulApin* was located on the left side of the Demon with Gaping Mouth (Cygnus) and thus might be identified as Equuleus although this identification is not certain. In some later sources this constellation was associated with the mythological bird Anzu and with the planet Mars. The distance between the constellations Equuleus and Orion are 2 x 72°, or about 144°; so once again the hypothetical system is affirmed. The next lover is the shepherd turned into a wolf. It may be an etiological explanation of the constellation Wolf, in some sources also associated with the planet Mars. The Wolf can be identified as the modern Triangulum. In many texts this constellation appears as the star of Anu, the god of heaven. Since Triangulum covers the area of about 72° back to Orion, it also fits the system of five constellations in which the heliacal settings of Venus can be observed in the 8-year cycle.

Number	Lover	Constellation	Modern name	Right ascension	Venus' cycle
1	Dumuzi	Shepherd	Orion	~50°	I 1
2	*allallu*–bird	?	?	~195°	I 5
3	the lion	Great Lion	Leo	~120°	I 3
4	the horse	Horse	Equuleus (?)	~265°	I 2
5	the wolf	Wolf	Triangulum	~335°	I 4
6	Išullanu	Abode of Anu (?)	Taurus/Gemini	~50°	II 1

Table 2. Lovers of Ishtar and their possible association with the 8-year cycle of Venus.

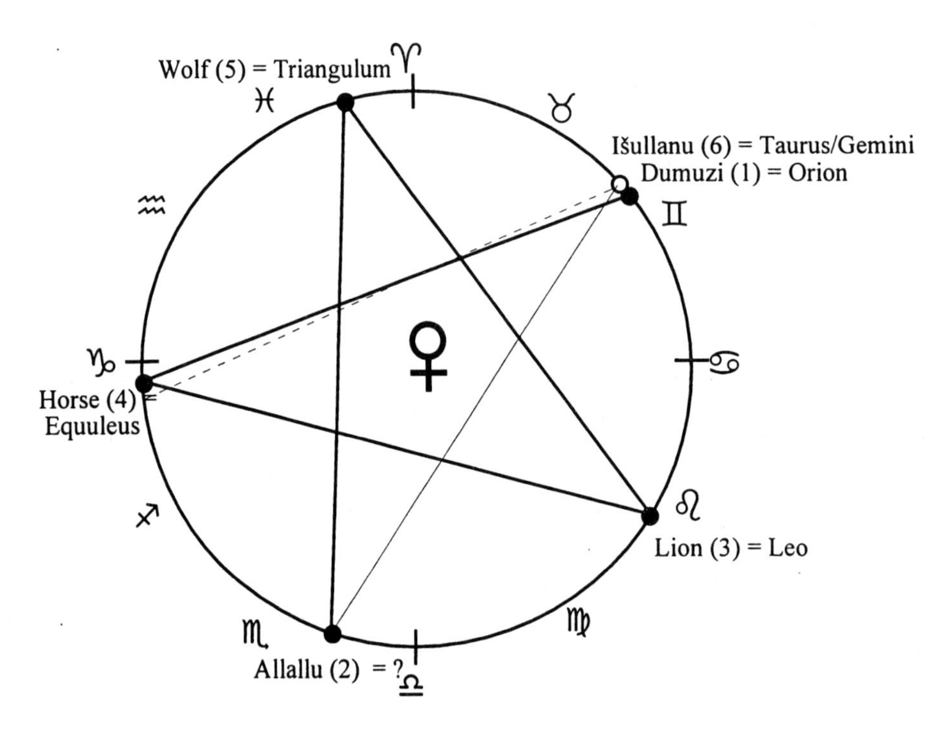

Fig. 1. Graphic presentation of the data in the Table 2.

The last lover on the list is Išullanu, the gardener of Anu. No Mesopotamian constellation is associated with this name, but its position suggests that it can be related to the region of the sky recognised as the abode of Anu, i.e. the surroundings of Taurus (Sołtysiak 2001). Thus Išullanu could be included in the system as the character for the place of internal conjunction of Venus in the next cycle. All data presented above are summarised in Table 2 and presented graphically in Fig. 1. The heliacal settings and risings of Venus in similar positions were observed in Babylon in the 7th century BC (Table 3) and, theoretically, could have been be observed also five earlier shifts of the 8-year cycle, i.e. in the middle of the 19th century BC.[3]

The sequence of the betrayed lovers of Ištar precisely corresponds to the sequence of constellations in which Venus heliacally sets in its 8-year cycle, under the condition that Orion is the starting point. Only the *allallu*–bird cannot be identified with a constellation. It should be noted that the sequence of lovers is not correlated with the time of heliacal settings, but with the right ascension

[3] The difference between eight solar years and five synodical cycles of Venus is about 2.44 days, which gives an average shift of 2.4° in the position of Venus at the beginning of two consecutive 8-year cycles. A shift of 72° occurs then each 240 years, and after such period the internal conjunction of Venus can be observed on the background of the same constellations. Although the sizes of particular constellations differ considerably, their sequence can be valid for a few 8-year cycles, i.e. for a few decades.

of constellations; also this time only the *allallu*–bird does not fit the system (Fig. 1). Of course all these correspondences may be due to chance. Taking into account the historical circumstances and the ritual context of the story, however, we cannot exclude the possibility that the list of Ištar's lovers was a commentary to the astral interpretation of the courtship of Ištar and Dumuzi, connecting it to all five internal conjunctions of Venus in its 8-year cycle and to all characters related to the constellations occuring as a background for these conjunctions. Of course it must be underlined that there is no direct proof of such an interpretation.

Year	Last visibility in the west				First visibility in the east			
	Julian date	Babylonian date	Right ascension	Constel-lation	Julian date	Babylonian date	Right ascension	Constel-lation
678 BC	May 21	Simanu 14	56°	Gemini	May 30	Simanu 23	51°	Orion
677/676 BC	Dec 29	Kislimu 10	276°	Capricorn	Jan 1	Kislimu 13	274°	Capricorn
675 BC	Jul 24	Du'uzu 22	128°	Leo	Aug 8	Abu 7	119°	Leo
673 BC	Mar 10	Addaru 25	348°	Piscis NE	Mar 12	Addaru 27	346°	Piscis NE
672 BC	Oct 2	Tišritu 7	203°	Scorpius	Oct 19	Tišritu 24	194°	Libra
670 BC	May 19	Simanu 10	54°	Ori/Gem	May 29	Simanu 20	48°	Orion

Table 3. Internal conjunctions of Venus in the years 678-670 BC, observed in Babylon (Parpola 1983: 391–392).

Finally let us return to the beginning of the story. In the opening passage Ištar invites Gilgameš to the house (which may be interpreted as heaven, in particular the abode of Anu) and offers him a chariot. In late esoteric Babylonian texts the constellation Chariot (modern Perseus, called also The Old Man) was associated with Venus since the sign for chariot resembles the sign U (interpreted homographically as Dilbat-Venus) in a box (George 1991: 161). Also the chariot of Marduk, in a bilinguial text from the late 2nd millennium BC, was called Dilbat in Sumerian and 'Ištar of the stars' in an Akkadian version (Lambert 1973: 279). Is that association exploited also in the story about Gilgameš? Unfortunately this question remains unanswered.

Acknowledgments
I wish to thank the Local Organizing Committee of the SEAC 2001 Conference (personally Mary Blomberg, Peter Blomberg, and Göran Henriksson) for financial support, which gave me the opportunity to take part in the conference. This paper was written within the project "Constellations in the mythologies of ancient Mesopotamia", financed by the Polish Committee for Scientific Research (Komitet Badañ Naukowych, grant nr 1 H01H 001 17). Last, but not least, I thank the Polish Science Foundation (Fundacja na Rzecz Nauki Polskiej), which awarded me a scholarship for the year 2001.

*Department of Historical Anthropology, Institute of Archaeology, Warsaw University, Krakowskie Przedmieście 16/28, 00-927 Warsaw, Poland, solar37@mail.uw.edu.pl.

References

Black, J. A., Cunningham, G., Fluckiger-Hawker, E, Robson, E., and Zólyomi, G. 1998. *The Electronic text corpus of Sumerian literature* (http://www-etcsl.orient.ox.ac.uk/). Oxford.

Campbell Thompson, R. 1924. The allalu–bird = Coracias garrulus Linn. *Journal of the Royal Asiatic Society of Great Britain and Ireland*, 258-259.

Dalley, S. 1989. *Myths from Mesopotamia: Creation, the Flood, Gilgamesh and others*. Oxford, New York.

George, A. R. 1991 Babylonian texts from the folios of Sidney Smith. Part Two: Prognostic and diagnostic omens, Tablet I, *Revue d'Assyriologie et d'Archéologie Orientale* **85**, 137-167.

Heimpel, W. 1982. A catalogue of Near Eastern Venus deities, *Syro-Mesopotamian Studies* **4/3**, 9–22.

Lambert, W. G. 1973. A new fragment from a list of antediluvian kings and Marduk's Chariot. In *Symbolae Biblicae et Mesopotamicae: Francisco Mario Theodoro de Liagre Böhl dedicatae*, ed. M. A. Beek, A. A. Kampman, C. Nijland, and J. Ryckmans, 271-280. Leiden.

Livingstone, A. 1986. *Mystical and mythological explanatory works of Assyrian and Babylonian scholars*. Oxford.

Parpola, S. 1983. *Letters from Assyrian scholars to the kings Esarhaddon and Assurbanipal. Part II: Commentary and appendices*, (Alter Orient und Altes Testament 5) Neukirchen-Vluyn.

Reiner, E. and Pingree, D. 1975. *Babylonian planetary omens. Part One. Enuma Anu Enlil tablet 63: The Venus tablet of Ammiṣaduqa*. Malibu.

Sołtysiak, A. 1999a. The Tree of Life and The Serpent of Truth: Celestial location and astronomical significance of the Paradise. In *Actes de la Vème Conférence Annuelle de la SEAC. Gdańsk 1997*, ed. A. Lebeuf and M. S. Ziółkowski, 41-67. Warsaw, Gdańsk.

Sołtysiak, A. 1999b. Adapa, Etana and Gilgameš: Three Sumerian rulers among the constellations. In *Oxford VI and SEAC 99. Astronomy and cultural diversity*, ed. C. Esteban and J. A. Belmonte, 289-293. Santa Cruz de Tenerife.

Sołtysiak, A. 2001 The Bull of Heaven in Mesopotamian sources. *Culture and Cosmos* **5:2**, 3-21.

Preliminary report on archaeoastronomical research in the Carpathian Basin during the Bronze Age

Emília Pásztor*

Abstract
There is good evidence that solar and lunar phenomena played a particularly important role in mythology in Europe in the Bronze Age. During that period the increasing use of special symbols assumed to be solar is well known and easily discernible on different types of archaeological artefacts. In the spread of bronze metallurgy another process must also be taken into account: the linked spread of a particular cosmology in which certain symbols might have had an important place. Were these real astral symbols, belonging to a certain archaeological culture or cultures, or were they so generally used that their origin cannot be determined? Could they symbolise a genuine sun or moon cult, i.e. something involving ritual practices that may be reflected in other aspects of the material culture?

Introduction
It is generally accepted that celestial phenomena influenced prehistoric people and that this influence might have been reflected or echoed in the material culture. The question is how to find the traces of these reflections and how to interpret them. I am in a very difficult situation when I want to study how celestial phenomena entered into and affected prehistoric peoples' ideas about their world in the Carpathian Basin. There are no standing stones, either in circles or rows; there are no megalithic tombs with their entrances turned towards presumably important directions; and there are no shrines built from earth or stones surviving as structures or ground plans for which I could try to find some deliberate directions and suppose about these directions that the sun or the moon or a constellation might have been their target. But the lack of these archaeological features does not mean automatically that prehistoric people in the Carpathian Basin did not care about the regular motions of the sun and moon or did not notice celestial phenomena in the sky. There are some promising hints in the latest research (Barlai and Bognár-Kutzián 2002). There are plenty of archaeological artefacts on which we can find motifs that are supposed to be solar or lunar symbols by the archaeologists. Human beings as symbol creators must have used symbols in prehistory as well (Renfrew and Scarre 1998). The purpose in using these signs or motifs cannot have been pure decoration, as there are also international symbols among them, symbols well known all over Europe and even outside Europe.

It is my firm conviction that in working with an archaeological problem that might involve some astronomical features we cannot avoid employing archaeological methods in the research, as pure astronomical interpretations can be misleading. Archaeoastronomical research can help cognitive archaeology in two ways:

1) by studying the symbols of time measuring and reckoning,
2) by studying peoples' relationship to the transcendental world.

However the archaeology of cult or religion is one of the most difficult areas of cognitive archaeology. Religious ideas cannot always be traced with unanimity in material culture. Evidence of cult activities is often difficult to recognise. Interpretations of symbols by modern researchers having altered ideas about the surrounding world are almost insoluble tasks.

There is good evidence that solar and lunar phenomena played a particularly important role in the mythology of Europe in the Bronze Age. Essays on prehistoric mythology frequently mention the evident existence of lunar or, especially, solar cults. Yet they invariably repeat a well-known list of examples of artefacts and rock art from many different parts of Europe. There have also been many studies of the orientations of Neolithic stone monuments. Yet no one has ever made a detailed, integrated investigation of tombs, artefacts, and other relevant archaeological evidence concerning the orientation in a particular area from the Neolithic onwards.

If we want to study the symbols used by the peoples of the Carpathian Basin during the Bronze Age, we should educe their possible origin and connections. In the Carpathian Basin the period from 2800 to 800 BC is considered the Bronze Age. According to the Hungarian archaeologists agreeing with the diffusion theory created by G. Childe, the strongest influences upon the Carpathian Basin consistently came from the southeast Mediterranean, including Greece (Childe 1950). They also assume that both in the Neolithic and in the Early and Middle Bronze Ages there were not only strong commercial connections between the two areas, but several occurrences of immigration. These people not only brought new economic knowledge, but also their beliefs and intellectual skills. Gordon Childe's followers consider this evident from numerous archaeological finds in the Carpathian Basin. They base the assumption mostly on echoes of the Minoan and Mycenean styles in decoration and rarely on forms of Bronze Age artefacts. The representation of the sun in Greek art can be traced with some certainty back to the Minoan and Mycenaean periods (Goodison 1989). Several depictions of solar symbols have been preserved in ritual scenes from those times. Among the most typical are examples to be seen on gold rings, but the 'realist' manner of representing the sun is also found in older works of art, for instance on seal rings, and it persisted through the Bronze Age period and early historical times.

Colin Renfrew, basing a new theory on the calibrated ^{14}C chronology (Renfrew 1973), presumes local, autochthonous development with a slight influence on the Balkans and the Carpathian Basin from the Eastern Mediterranean during early prehistory. He even assumes early independent metallurgical centres in the Balkans that might have influenced northern Greece. This is the possible historical background of the Carpathian Basin during the Bronze Age.

When studying cult activities in the Carpathian Basin, we are confronted with another difficulty. There are many more remains of such activities in the Balkans, as well as in the Aegean, than in the Carpathian Basin. These activities might have been performed by special persons designated for this role in special enclosures in the Balkans and eastern Mediterranean. Unfortunately, in the Carpathian Basin during the entire Bronze Age there were no places for legal/formal religious life. According to the excavations the cult objects come from cemeteries, graves, or from pits in the living area of a settlement. This means that cult activity must have been performed in the family circle or inside a settlement. Even in an archaeological culture it is not sure that we can talk about uniform cult activity with common disciplines or rules. Supposedly, the ceremony for the fertility of a community was performed by the community itself; the people themselves made or created the necessities as well. This can be the reason or explanation for the many kinds of astral symbols, if they are astral symbols at all. However there is a slight indication, although dim, of the existence of astral symbols. There are abstract signs that are international and believed to be astral or sun symbols, and these can be hunted out among the decorations on Bronze Age artefacts.

As a first step I have been trying to collect the international symbols that are among the patterns covering the surface of artefacts (Chevalier and Gheerbrant 1973). The so-called swastika belongs to these.

Swastika

The swastika is a very old ideogram. The first such signs preserved to our time were found in the Euphrates-Tigris valley and in some areas of the Indus valley. They seem to be more than 3,000 years old. A large number of seals discovered in Harappa, one of the prime sites, bear swastika designs. A swastika is supposed to be of two types:

1) The right-handed or male, representing the vernal sun and the god Ganesha. In this swastika the extremities of the arms of the cross bend clockwise. It is considered auspicious.
2) The left-handed or female cross represents the autumnal sun and the goddess Kali. Its arms are bent in an anti-clockwise direction and it is believed to be inauspicious.

In the Vedas it is referred to as the wheel of the sun. It indicates cosmic procession and evolution around a centre.

The Sumerians seem to have used the swastika, but none of their successors seems to have preferred to apply it, as compared to other ancient Eurasian cultures. Count Goblet d'Alviella (d'Alviella 1979 [1894]), who at the end of the nineteenth century conducted research in the distribution and migration of sacred symbols, put forth the theory that certain symbols were mutually exclusive, i.e. they could not appear in the same country or cultural sphere. According to this theory the swastika and the round disc with horizontally outspread wings, the circle with the four-pointed star, and the four-armed cross in a circle are all symbols for the sun, the highest god and the supreme power and life force. The swastika became commonly used in Greece from around 700 BC and was freely used in antiquity for decorations on ceramic pots, vases, coins, and buildings.

In the Carpathian Basin the swastika occurs first in an archaeological group called the Tisza culture as early as the Neolithic Age. Two of them straddle a human face on the neck of a so-called face-pot

(Csalog 1959: fig. 4). In the Bronze Age it continues to emerge among the ornaments on urns (Vicze 2001), small figurines (Kovács 1991), and on bronze weapons—possible votive offerings (Mozsolics 1967, Mozsolics 1973).

The swastika's spectrum of meaning is centered about power, energy, and migration.

Cross with equal arms

The cross also seems to have been associated with the sun and the powers that control the weather; this assumption can be read in many places. In Babylon the equal-arm cross was considered one of the attributes of Anu, god of the heavens. In the mighty Assyrian empire, which seems to have originated as a Babylonian colony in the second millennium BC, the sun cross in the wheel-cross form was one of the attributes of the national god Assur. Apart from rock carvings from the Neolithic Age the first cross structures were those of the wheel cross. On the other hand, it may also be a graphic representation of the centralizing effort of the social structure of ancient societies, which was associated with the invention of the wheel, as the meaning *town* of the Egyptian hieroglyph suggests.

The archaeological finds from the Carpathian Basin of the Bronze Age are often decorated with this sign and its wheel-cross version as well. I mention only the most well-known representation of them, the pendants of the Koszideri treasure, which are cross-wheels cast in bronze (Mozsolics 1967: Tafel 46). As this type of symbol can also be found on the bottom of the vessels of later periods, an important role is assigned to them in fertility rites (Sági 1962).

Crosses with one longer arm are considered symbols with different meaning.

Half spirals in clockwise and anti-clockwise directions

The spiral in clockwise direction most probably represents the sun, but it might also have been used for recurring migrations or tribal wanderings. In Viking-age rock engravings and paintings found in Sweden, it is often used to mean potential movement or independent movement (against the sun, waves, and wind when necessary) and eventual return. In ancient Greece a similar structure was used to represent the zodiacal sign Leo. This zodiac sign is strongly associated with potential power, strength, etc.

Valcamonica

There are rock carvings all over Europe, except in Hungary. Searching for possible interpretations or roles of symbols in ritual rock carvings can help very much towards the understanding of offering groups of symbols or signs. The rock-carving site nearest the Carpathian Basin is in the Valcamonica valley.

The first images of the sun in the rock art of Valcamonica are engraved on the menhirs and boulders of the Copper Age. In the period 2800-2400 BC, the sun is associated with weapons and animals, probably to symbolise a male god. In the next phase, parallel to the chronological horizon of the Bell-Beaker period of 2400-2200 BC, it is positioned—like a crown—above the head of a male anthropomorphic figure, which is sometimes associated with weapons, animals, and two other figures—one male and the other female. The crowned figure is interpreted as the anthropomorphisation of the sun god, armed, of the previous period. This last phase is richly represented in Valcamonica.

In the Bronze Age rock art of Valcamonica the sun figure is rarely represented with sunbeams, compared to the previous period. It is symbolised by a disk with a central point that is frequently associated with praying figures: The radiate wheel appears during the Bronze Age of the Valcamonica. It is interpreted sometimes as the representation of the praying figures associated with the sun (rock 50 of Naquane, rock 49 of Luine), or as a wheel of the chariot and, in this sense, as a symbol of the chariot of the sun that in the ancient legends transports the sun, drawn by water birds. In other cases it can be interpreted as the representation of a shield in association with other images of weapons such as daggers, halberds, and axes (Anati 1994).

Preliminary conclusions and emerging questions

Although my work is still in the phase of collecting data, I can frame some possible general rules and further questions:

- In the Carpathian Basin the different archaeological groups called cultures use these symbols in different ways. Those groups that show most southeast influence seem to favour these symbols.

- Bronze metallurgy in the Carpathian Basin took off during the middle period because of the nearby Transylvanian mountains, which were rich in ore and have a central position in this part of Europe. The early simple tools made from bronze sheet are followed by richly decorated heavy bronze weapons, tools, and priceless bronze and gold jewellery. Their surfaces abound with special motifs and signs, including sun symbols. Most of the finds come from votive offerings of treasures; 323 items of bronze and 21 items of gold have been found. There are many axes and sickles in these finds, which were used in making the earth fertile. Were they devoted to the sun god? Among the weapons there are numerous examples that must have been symbolical ones, judging from their size or quality. The sun symbols on them might have protected their owners and/or represented their power.
- From the later examples, for instance from the Medieval Period, we know that these symbols had protective power.
- According to the examples of southeast Europe we can suppose that the ceremonies in connection with vegetation took place around flames of the fire in the middle of the houses, where the votive offerings of small amounts of the crop were placed in human or animal-shaped pots. The fireplace itself was the object of cult activity. It was an altar belonging to the god who was emanating heat, like the sun. That is why there are sun signs on the fire equipment, such as firedogs (Nagy 1979). In the Carpathian Basin the numbers of these tools had increased by the final period of the Bronze Age. They have been found mostly in settlements, which prove their home-altar significance.
- In a cemetery of the same period very special rather strange small finds came to light. The excavator unearthed small clay objects that represented the sun, moon and stars in different forms. For example: plain sun or moon disc, sun disc with rays, moving sun symbolised by the wheeled cross, crescent. These were piled in and under urns. Each of them had a small handle. Might they have represented the whole sky with its light sources, with gods and goddesses, or certain phases of these celestial objects, which they needed for their fertility cult? Among the discs there is one with a single ray coming from it. Does it symbolise the first ray of the spring sun, which is a separate god in several religions (Nagy 1979: 66)?

Acknowledgement
I am grateful to the Foundation for Hungarian Science, which supports this research.

*Foundation for Hungarian Science, Dunafoldvar, Sohaz u. 4 H-7020, Hungary, karath@enternet.hu.

References

d'Alviella, G. 1979. *The migration of symbols*. London (facsimile of the 1894 edition).
Anati, E.. 1994. *Valcamonica rock art: A new history for Europe* (Studi Camuni 13). Capo di Ponte.
Barlai, K. and Bognár-Kutzián, I. 2002. *"Unwritten Messages" from the Carpathian Basin*. Budapest.
Chevalier, J. and Gheerbrant, A. 1973. *Dictionnaire des symboles*, 6th ed. Paris
Childe, V. G. 1950. *Prehistoric migrations in Europe*. Oslo.
Csalog, J. 1959. Die anthropomorphen Gefässen und idolplastiken von Szegvár-Tűzköves. *Acta Archaeologica Academiae Scientiarum Hungaricae* **11**, 7-38.
Goodison, L. 1989. *Death, woman and the sun*. (Bulletin of the Institute of Classical Studies, Supplement 53). London.
Kovács, T. 1991. Kis figurák. *Folia Archaeologica* **41**, 7-16.
Mozsolics, A. 1967. *Bronzefunde des Karpatenbecken*. Budapest.
Mozsolics, A. 1973. *Bronze- und Goldfunde des Karpatenbeckens*. Budapest.
Nagy, L. 1979. A tűzikutya és a holdidol kérdése a magyarországi leletek alapján. *Veszprém Megyei Múzeumok Közleménye* **14**, *19-69.*
Renfrew, C. 1973. *Before civilization: The radiocarbon revolution and prehistoricEurope*. London.
Renfrew, C. and Scarre, C. (ed.). 1998. *Cognition and material culture: The archaeology of symbolic storage*. Cambridge.
Sági, K. 1962. Árpád-kori varázslás régészeti emlékei. *Veszprém Megyei Múzeumok Közleménye* **3**, 55-85.
Vicze, M. 2001. The symbolic meaning of urn 715. *Komáron-Esztergom Megyei Múzeumok Közleménye* **7**, 119-133.

Possible astronomical orientation of the Dutch *hunebedden*

César González-García* and Lourdes Costa-Ferrer**

Abstract
From about 3400 to 2850 BC, the peoples of the Western Group of the Funnel Beaker Culture occupied the North European plain between Amsterdam and Hamburg. In the northern Dutch provinces of Drenthe and Groningen, running along the 'Hondsrug', lie 52 passage graves belonging to this culture. The monuments consist of a number of trilithons that form the burial chamber. For some of the chambers there is a passage that, although not very long on one side, also consists of a series of trilithons. An earthen mound originally covered all of the stones. Archaeological studies in the past roughly determined an east-west orientation for the chambers, indicating that the passages are always on the east or south sides of the chambers. However few more detailed studies are found in the literature.

We have measured the orientation of both the chamber and the passage of each grave wherever this is possible. The chamber orientations are broadly distributed about east-west with a peak near the equinoxes. It has already been proposed that they may have incorporated deliberate solar and lunar alignments. The orientations of the passages (entrances) exhibit a slightly broader distribution. This is centred about due south, but is double peaked, with a peak on either side of South. One might postulate an alignment to the culminating moon low in the southern sky at times around the major standstill. Alternatively the data could be interpreted as relating to the rising and setting of a group of stars, notably Alpha Centauri and perhaps the Southern Cross.

Introduction
In the northern Dutch provinces of Drenthe and Groningen there is a cluster of Neolithic passage graves belonging to the *Tricherrandbecher* (TRB) or Funnel Beaker Culture. The passage graves that we shall discuss belong to the Western Group, the remainder of which is found in western Germany. The other groups are the Altmark Group in Northern Germany and the North Group, which reaches southern Scandinavia. In the region of interest in this paper the TRB Culture is estimated to date from about 3400-2850 cal BC (Bakker 1992; Baldia 1995). Although there is no good archaeological reason for considering the Dutch group in isolation from the remainder of the Western Group, we shall do so for the purposes of this preliminary study, leaving for the near future a comparative study with data from the German group.

There are 76 known sites of passage graves in the Netherlands, although some are now destroyed (Bakker 1992). Fifty-three sites are sufficiently well preserved for measurements to be made, although one is not now in its original place. In a flat and sandy country such as the Netherlands it is perhaps surprising to encounter megalithic monuments with stones up to a tonne in weight. However they are all located in an area known as the 'Hondsrug', or Dog's Back, a low ridge running from the city of Groningen in the north to Emmen in the south (Fig. 1). This ridge was created as a result of glacial action, which was also responsible for depositing a number of large erratic boulders.

These monuments are known in the Netherlands as the *hunebedden*, or 'giants' beds'. They have been subject to spoliation from their very beginnings, in some cases dating from Neolithic times. During the Middle Ages their stones were used in the building of churches, some of which may still be seen in the villages of the area. In 1734-1735, the Province of Drenthe made a law to protect these monuments, which included a 100-golden-gilders fine, a good amount of money for that time. This law, which can be considered as only the second in Europe passed to protect archaeological sites, was necessary due to the plundering of the boulders for the construction of dams, as wooden dams where found to be less reliable. In the 19th century, the state and the province of Drenthe began to buy the fields with the *hunebedden* with the intention of protecting them, as the 1734 law was no longer effective by that time. Afterwards there was some restoration work. According to a contemporary hypothesis the tumuli were believed to be dunes that were placed there by the action of time, covering the true structure of the monuments. Thus a major restoration uncovered the stones from their original mound. And this is the aspect they present today.

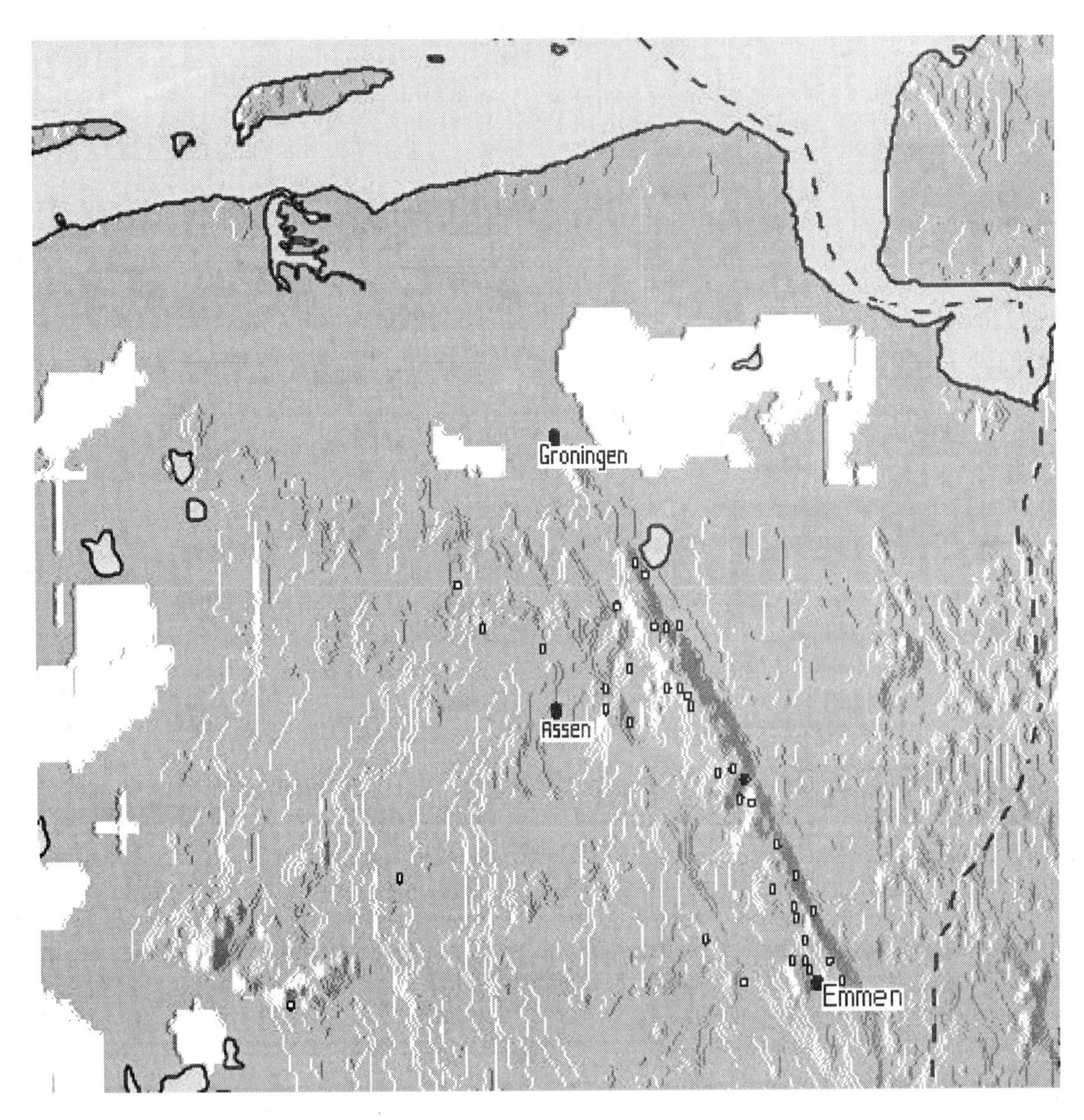

The first accurate plans and profiles of *hunebedden* were made by Sir Henry Dryden during the expedition of the Society of Antiquaries of London in 1878; however they were never published. Copies of them are kept in London, Oxford, and Assen. W. J. de Wilde documented the *hunebedden* in 1904-1906. J. H. Holwerda started modern *hunebed* excavations in Drenthe in the 20th century (D19-D20-D43). A. E. van Giffen continued the excavations (D53, D40, D30, D21, D22, D26, D16, D39, as well as some demolished *hunebedden*). The results from these excavations were published together with plans and figures in *De hunebedden in Nederland*, (van Giffin 1925-27) in two volumes and one atlas. Later, van Giffen was involved in the conservation and restoration of the monuments, in some cases even at some risk. For example the *hunebed* D53 in Havelte, one of the largest in the Netherlands, had to be demolished during the Second World War to build an airfield, on German orders. Van Giffen marked all the stones and the sites and, at the end of the war, carefully put them all back in place.

The extent of conservation of many of the *hunebedden* before van Giffen's actions was small. However he managed to carefully recover many of the sites of the stones during the decades of the 1950's and 1960's. Today most of them have undergone major restoration (lifting up of stones) after excavation (in recent years by Bakker in 1983, De Groot in 1988, and others; for a recent review see Bakker 1992). This is another important caveat to keep in mind, as it may be a source of possible errors in some of the measurements.

The dating of these monuments comes from only 15 reliable radiocarbon dates for this group and gives the above-mentioned 3400-2850 cal BC.

From these excavations and restoration works several fragments of pottery were recovered. Many of them can be seen at the Drenthe Museum in Assen (Fig. 2).

Figure 2. Pottery of the Funnel Beaker Culture.

Many typologies have been debated by various authors, and conflicting estimates have been given for the duration of the different phases; Bakker (1992) accounts for 10 phases. Few other remains have been collected from the chamber fill. The soils in the area are too acid, and few human or animal remains have survived. Offerings have been recovered at the entrance to the tombs, similar to those found in Scandinavia (Hård & Roslund 1991).

Several theories have been proposed to explain how these monuments appeared in their culture and the role played by these burial places in the TRB society. As to the origin a western one (through Germany and France) and an eastern one (through the Chech Republic and Poland) have been put forward, but an autochthonous origin seems to be favoured (see Baldia 1995: chap. 5 for a recent review).

The role played by these monuments is also a matter of debate. It is unclear as to whether they were central places or territorial markers, or whether they defined the territory of a chiefdom (Baldia 1995: chap 15). Baldia (1994), for example, gives the idea that they could have been used as route markers, facilitating communication and trade. Whatever their role it seems clear that a high degree of technological skill along with manpower was needed to build these monuments, which speaks for a complex social organization. The existence of megaliths may imply the religion of ancestor worship, giving a cohesive meaning to the community building them.

Once we start to explore their possible meanings, the question of their orientation cannot be ignored (Hoskin 2001). If they conformed to a common ritual tradition or if seasonal factors influenced their construction or use, then certain patterns in their orientations might be expected. This is the possibility we investigate here.

The sample

Of the 56 catalogued *hunebedden* by van Giffen, one (Grand Dolmen G5, Baldia 1995) was moved to the local museum in Delfzijl shortly after its discovery in the early 1980's to save it from destruction during the construction of a factory. In Table 1 we give our measurements of the orientations of 52 *hunebedden*. No data are given for three sites: D33 was in such bad condition that van Giffen used some of its stones to rebuild D34 nearby; D44 (uniquely) is still on private land and badly damaged; and D48 is a huge single slab and not a *hunebed* as originally supposed by van Giffen.

In the table we also give the type of passage grave, following Bakker's classification (1992): *N* denotes a simple grave without a surrounding kerb, *P* a grave with a kerb, and *Q* a so-called 'langbett', a long-grave with a kerb enclosing more than one passage grave. There is only one type-Q grave in the Dutch group, although others can be found in neighbouring Germany. The numeral accompanying the type designation (N0, N1, P2, etc.) indicates the number of trilithons marking the entrance.

Orientations were measured using a Suunto compass-inclinometer and they were measured from both ends of the graves. Instrumental (random) errors are estimated to be ± 1.0° and this is indicated at the head of the table. Systematic errors due to positioning during the measurement process are estimated to be no greater than ± 0.3°. In some cases, for example where the *hunebed* chamber is not straight (possibly due to errors in the reconstruction, or to shifting of the stones over time, or perhaps as originally intended), an estimate of the larger error introduced is given. Where it was feasible to measure the orientation of the entrance, this is also given. In all but two cases the horizon altitude was zero to within our inclinometer's error, i.e. ± 1.0°, the two exceptions being the horizon in the direction of the entrance orientations at D53 and D54 (2° and 3° respectively). Measurements were corrected for magnetic declination. The last two columns give the geographical coordinates, measured

Hunebed	Type	Chamber azimuth ±1°	Passage azimuth ±1°	Latitude	Longitude
D1	N1	79	161	*53 6	6 25
D2	N1	84	163	*53 4	6.27
D3	N1	99	192	53 6 27	6 40 21
D4	N1	90	185	53 6 27	6 40 21
D5	N1	106	194	*53 3	6 32
D6	N0	121		*53 5	6 38
D7	N0	81	167	53 3 41.6	6 41
D8	N0	73		53 3 36.7	6 42
D9	N1	85	184	*53 4	6 43
D10	N0	104		*53 2	6 39
D11	N1	112	193	*53 1	6.42
D12	N0	146		*53 1	6 43
D13	N0	69	168	53 0 41	6 43 37
D14	P2	99±2	182	53 00 07	6 43 51
D15	P2	112	194	*53 1	6.37
D16	N1	82±2	177	*53 0	6 37
D17	N1	108	206	52 59 23.5	6 38 58.9
D18	N1	87	178	52 59 25.6	6 38 59.9
D19	N2	119	205	52 57 8.3	6 47 9.2
D20	P2	96	184	52 57 8.1	6 47 10.6
D21	N1-0	46±2		52 56 40.2	6 48 3
D22	N0	73.5		52 56 40.7	6 48 3.4
D23	N0	95.5±8		52 56 40.8	6 48 19.7
D24	N0	89.5		52 56 40.8	6 48 19.7
D25	N0	72.5		52 56 40.6	6 48 19.7
D26	P2	66	155	*52 57	6 46
D27	N2	109.5	201	52 55 48	6 47 48
D28	N0	99		52 55 34.4	6 48 42.2
D29	N1	71	162	52 53 33.5	6 48 43.2
D30	N1	170.5	78.5	52 53 30.5	6 50 47
D31	N1	119	202	52 52 5	6 52 16.6
D32	N0	68		52 51 25.3	6 50 23.7
D34	N1	138.5	230	52 50 34.3	6 52 5
D35	N0	89.5		52 50 7.7	6 52 16.7
D36	N1	104.5	193	52 50 24.1	6 53 42.5
D37	N0	99		52 50 24	6 53 43.6
D38	N1	44	143	*52 49	6 53
D39	N0	25		*52 49	6 53
D40	N1	163.5	77	*52 49	6 53
D41	N0	72.5		*52 48	6 53
D42	N0	129.5±2.5		*52 48	6 53
D43N	Q1	17.5	98	52 47 36.2	6 53 15.2
D43S	Q1	17.5	99	52 47 36.2	6 53 15.2
D45	P2	77.5	174	*52 48	6 55
D46	N0	96		*52 47	6 56
D47	N0	159.5		*52 47	6 56
D49	P2	117	215	*52 49	6 45
D50	P2	87.5±2.5	177	*52 47	6 48
D51	N2	72.5	160	*52 47	6 48
D52	N1	55	152	*52 52	6 20
D53	P2	74	163	*52 48	6 13
D54	N1	112	197	*52 48	6 13
G1	N2	120±4	210	53 6 57	6 39 30

Column 1
The *hunebed* code, following van Giffen (1927).

Column 2
The architectonic type (see text).

Columns 3 and 4
The measured azimuth in degrees, ± 1°.

Columns 5 and 6
The geographical coordinates.

No measurements for D33, D44, D48, and G5, see text.

Table 1. Measurement data for *hunebedden*.

Fig. 3. Two examples of *hunebedden*. Left, D17-Rolde is catalogued as N1. Right, D50-Noordsleen is catalogued as P2, note the surrounding kerb.

with a GPS. Where this was not possible, the values have been taken from the literature and are marked with an asterisk.

Location and orientation

In Fig. 1 we show the distribution of the *hunebedden* in the landscape. As already mentioned, they are located within a broad band joining Groningen and Emmen on land that is somewhat higher than the surroundings. In Neolithic times it would probably have been surrounded by marshes (Bakker 1976). The mean elevation of this area is 12-15 m above sea level. Baldia (1994) has suggested that they are clustered along a Neolithic route.

Van Giffen (1927) stated that most of the *hunebedden* have an east-west orientation, but he did not make a statistical study. Baldia (1995), who collected from the literature the measurements of 2385 chambers from all over the TRB Culture area, also found no clear pattern. But considering all the monuments together would fail to reveal any practices specific to a particular locality and, furthermore, he divided the circle into just 16 sectors, too course a bin size to reveal any precise trends in the data.

If we plot the mean azimuths of the *hunebedden* (Fig. 4), we can see immediately that there is a clear general trend for east-west, rather than north-south, orientations. We also see a broad concentration around due east (mean = 94°, standard deviation = 30°). A large proportion of the chamber orientations lie between the solar (83%) and lunar (88%) extreme positions. On the other hand, while Reijs (1997) has claimed more specific alignments to particular rising and setting points of the sun and moon, such as full moonrise immediately before the equinox, the same author finds this explanation to be inconsistent, and we find no data to support these more specific alignments. Instead, the pattern of orientations is fully consistent with the sun rising/climbing/descending/setting model proposed by Hoskin (2001).

When the passage graves were built, the chambers were covered with an earthen mound that in many cases was elliptical or kidney shaped, with the mound being added after the chamber had been completed. Admittedly the chamber could have been oriented towards the sun or moon on the day of construction, and orientation could not then have been used after the event, but perhaps the purpose was simply to get things right at the time of construction. The entrance, however, was open and, as previously mentioned, is always located on the eastern or southern side of the barrow.

Since the chambers are in most cases oriented east-west, the entrances would have to be either to the south or north. They are located mostly to the south, which would be in accordance with the avoidance of northerly directions proposed by Hoskin. Offerings were made at the entrance after construction of the monument (Bakker 1992), indicating a ritual practice there.

Baldia (1995) records the orientations of 786 chamber entrances and notes that they are predominantly to the south, with two clear peaks: one due south and the other close to east. There are also indications in other local groups of TRB tombs that the entrance was cardinally oriented (Hårdh & Roslund 1991, for the Scandinavian group). Baldia found, in addition, that the orientation of the entrances is not necessarily perpendicular to the gallery; in many cases they deviate from perpendicularity by several degrees. There is no clear reason for this, although one possibility is that

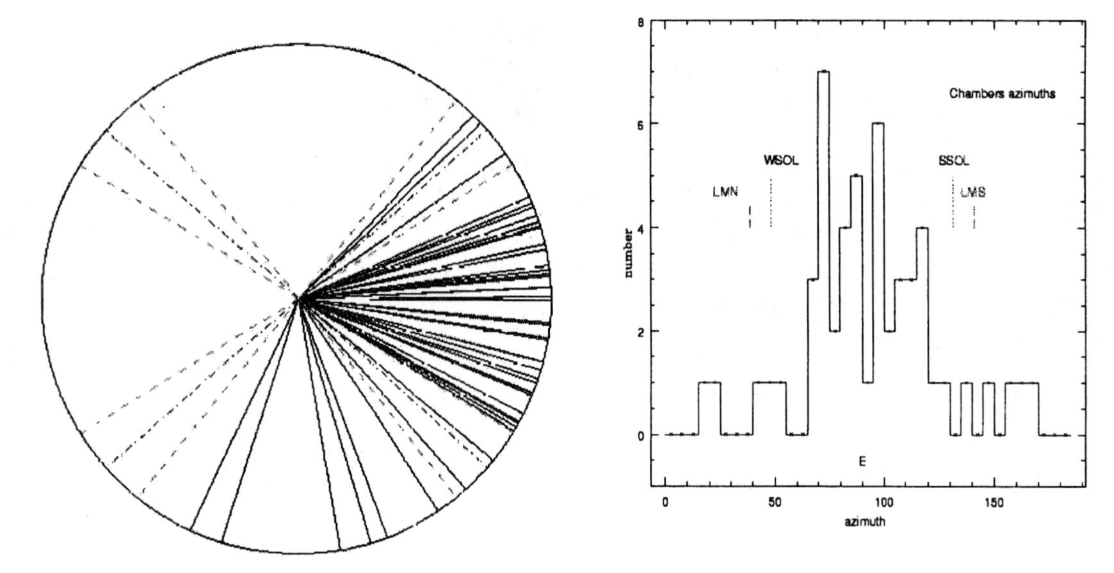

Fig. 4. Azimuths of the chambers. Most of them lie within the extreme positions of sun and moon. The left figure shows the eastern azimuth for each chamber. Dashed lines indicate the extreme positions of the moon while dotted-dashed lines indicate the solstices. The right figure shows a histogram with the same results.

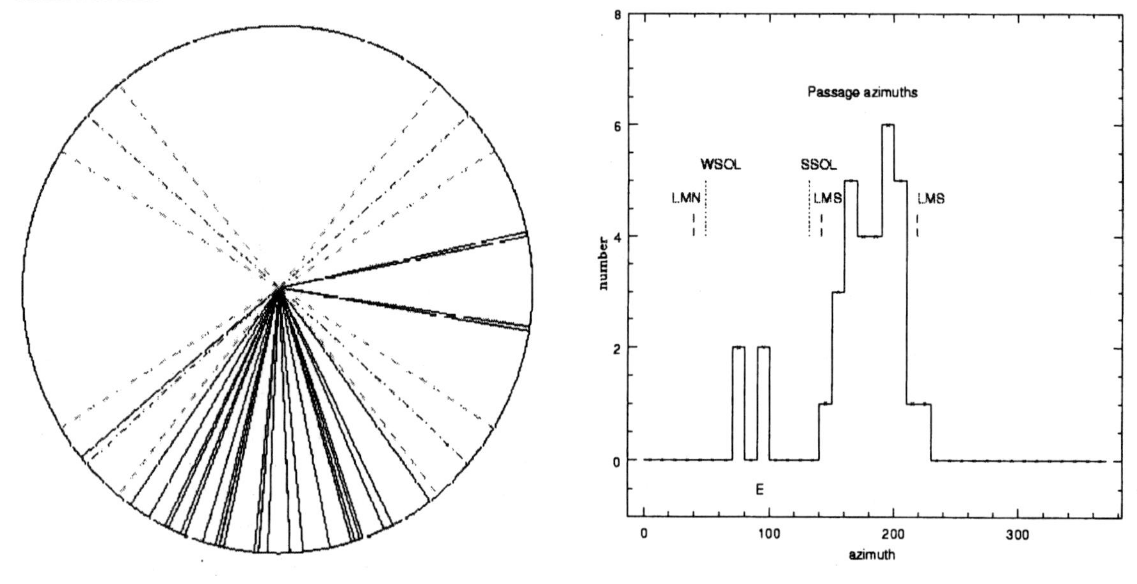

Fig. 5. Azimuths of the passages, where available. Notice the double-peaked distribution on either side of due south.

the entrance was oriented towards something else (Baldia, private communication). Given the lack of prominent topographic features in the area (such as mountains or river valleys), we should check for possible astronomical alignments. With this in mind we plot the distribution of azimuths of the passage entrances in Fig. 5. In the histogram we see a clear concentration around the south, with some suggestion of a double peak. This may indicate a pattern of alignment towards the rising and setting of particular celestial bodies.

If we transform azimuths to declinations, we find a strong peak around declination –35° (Fig. 6), which is largely to be expected for a set of orientations concentrated around south at latitude 53° (co-latitude 37°). If we suppose that the passages were oriented towards celestial objects, one possibility could be the moon close to its southernmost extreme. The full moon near the summer solstice would pass the meridian at a low altitude—no more than 7° in major standstill years—and could be a spectacular phenomenon, as noted by Burl (1980) and Ruggles & Burl (1985) for the recumbent stone circles in Scotland (where the moon would pass even closer to the southern horizon). The southernmost declination of the moon (–30°) is marked on the plot as LMS. The main peak is at a lower declination than this, and indeed we can see in Fig. 5 that most of the passages lie within the rising and setting of the moon at the major standstill limit. This means that if the moon were the target, then it would have to have been so when it was up in the sky rather than rising or setting. A full moon low in the sky could certainly have lit up the entrance of a passage grave. On the other hand, in years well away from the major standstill, the midsummer full moon could reach altitudes as high as 17° at the meridian, and perhaps the need to orient passages upon a low moon, i.e. either side of due south, could help to explain the double peak in the azimuth distribution.

Another possibility is that of a stellar alignment, although demonstrating convincingly that such alignments were deliberate is problematic, as it is usually possible to fit at least one bright star to any alignment by choosing an appropriate date (Ruggles 1999: 52). If we look at the sky at the time of the dates for these monuments, i.e. 3400 to 2850 BC, we find several bright candidates between declinations –40° and –30°. We have chosen Alpha Centauti, Beta Centauri, the asterism of the Southern Cross, and Beta Canis Majoris—all of them brighter than magnitude 2. Outside this declination interval we have included Sirius and Rigel for comparison. All these stars would have been visible at the time.

In Fig. 6 we indicate these stars and their declinations during the period 3400 to 2850 BC with a dark arrow. We also give their present magnitude. If we compare the declinations of the passages and those of the stars, we see a strong pattern of correlation between the entrance orientations and the six highly conspicuous stars in the southern sky, the Southern Cross, and the Pointers (Alpha and Beta Centauri).

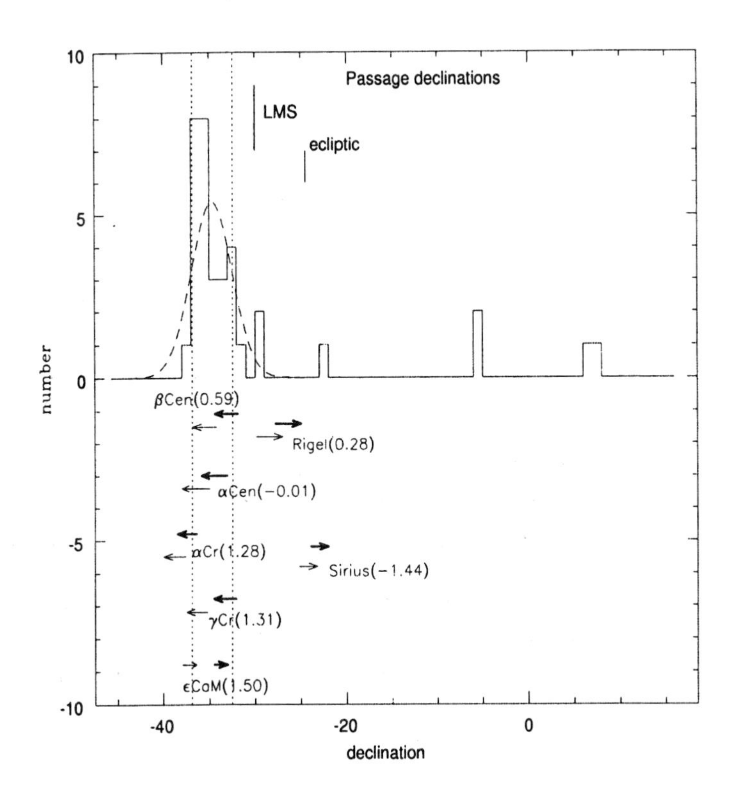

Fig. 6. Declinations for the passages. There is a peak around declination –35°. The declinations of the ecliptic and the lunar major stanstill are also shown. In the lower part of the plot the declinations of some bright stars are given for the 500-year period (dark arrows) as well as the declinations when atmospheric extinction is taken into account (light arrows). For details, see text.

This is not the first time a stellar alignment with these candidates has been claimed (Hoskin 1989; Fodera Serio et al. 1992; North 1996; García Rossello et al. 2000; Esteban et al. 2001). The biggest problem with these alignments is that these stars would have been quite low above the horizon, reaching an altitude of only about 7° at most. Atmospheric extinction would render even these stars invisible at the horizon (Schaefer 1986), although they would be conspicuous as they passed from east to west across the meridian. A movement of the declination of the passages would reflect this. We have indicated the shift in Fig. 6 with light arrows for each star. This shift depends, as noted above, on the magnitude of the star and can be important, as we can see. In the same figure we have fitted the peak around declination -34° to a Gaussian. Vertical dotted lines give us the 1-sigma region. In the light of this only the Pointers would have been visible for the whole 550 years. However these are only the brightest stars in that region. To know if they were really intended as targets, for whatever reason, we would need further ethnoastronomical information, which we lack today.

Acknowledgements
The authors wish to thank Juan Antonio Belmonte and Clive Ruggles for all their support and useful comments in the preparation of this text. We are also grateful to Jorge Jimenez for stimulating discussions, to Max Baldia for making available his PhD thesis as well as for maintaining a fruitful dialogue, and Jacquie Keane for carefully reading and correcting our English.

*Kapteyn Astronomical Institute, Groningen, The Netherlands, cesar@astro.rug.nl.
**Kapteyn Astronomical Institute, Groningen, The Netherlands, lourdes@astro.rug.nl.

References

Bakker, J. A. 1976. On the possibility of reconstructing roads from the TRB period. *Berichten van de Rijksdienst voor het Oudheidkundig Bodemonderzoek* **26**, 63-91.

Bakker, J. A. 1992. *The Dutch hunebedden: Megalithic tombs of the Funnel Beaker Culture* (International Monographs in Prehistory, Archaeological Series 2). Ann Arbor.

Baldia, M. O. Megalithic tombs and interregional communication. Unpublished paper presented at the international symposium *Megaliths and Social Geography*, 13-17, May 1994, Falköping, Sweden. Available at the web site http://www.comp-archaeology.org/Falbytxt.htm.

Baldia, M. O. 1995. A spatial analysis of megalithic tombs (Dissertation presented to the graduate faculty of Dedman College, Southern Methodist University, Dallas, Texas).

Burl, A. 1980. Science or symbolism: Problems of archaeo-astronomy. *Antiquity* **54**, 191-200.

Esteban C., Belmonte J. A., Perera Betancort, M. A., Marrero R. and Jiménez González, J. J. 2001. Orientations of pre-Islamic temples of northwest Africa. *Archaeoastronomy* (Supplement to the *Journal for the History of Astronomy*) **26**, S65-S84.

Fodera, S. G., Hoskin M., and Ventura F. 1992. The orientation of the temples of Malta. *Journal for the History of Astronomy* **23**, 107.

García Rosselló, J., Bisquerra, J F., and Hoskin M. 2000. Orientations of the Talayotic sancturies of Mallorca. *Archaeoastronomy* (Supplement to the *Journal for the History of Astronomy*) **25**, S58-S64.

van Giffen, A. E. 1925-1927. *De hunebedden in Nederland*, 2 vols and an atlas. Utrecht.

Hårdh, B. and Roslund, C. 1991. Passage graves and the passage of the moon. In *Regions and reflections: In honour of Märta Strömberg* (Acta Archaeologica Lundensia, series in 8°, 20), ed. K. Jennbert, L. Larsson, R. Petré, and B. Wyszomirska-Werbart, 35-43. Lund.

Hoskin, M. 1989. The Orientations of the Taulas of Menorca (1): The Southern Taulas. *Archaeoastronomy* (Supplement to the Journal for the History of Astronomy) **14**, S117-S136.

Hoskin, M. 2001, *Tombs, temples and their orientations*: *A new perspective on Mediterranean prehistory*. Bognor Regis.

North J. 1996. *Stonehenge: Neolithic man and the cosmos*. London.

Reijs V. 1997. http://www.iol.ie/~geniet/eng/.

Ruggles, C. L. N. and Burl, H. A. W. 1985. A new study of the Aberdeenshire recumbent stone circles, 2: Interpretation. *Archaeoastronomy* (Supplement to the *Journal for the History of Astronomy*) **8**, S25-S60.

Ruggles C. L. N. 1999. *Astronomy in Prehistoric Britain and Ireland.* New Haven.

Schaeffer B. E. 1986. Atmospheric extinction effects on stellar alignments. *Archaeoastronomy* (Supplement to the *Journal for the History of Astronomy*) **10**, S32-S42.

Astronomical aspects of megalithic monuments in Siberia

Leonid Marsadolov*

Abstract
Megalithic monuments are known in both western Europe and Siberia. The Great Salbyk barrow is the best known of the type in Siberia. The barrow height is more than 20 m and originally it was pyramid-shaped. Under the mound was a square 'fence' (70 x 70 m) made of huge stone slabs placed vertically and horizontally and weighting some tonnes (the average size was about 5 m wide). Inside the fence a square pit-grave had been dug, and there were seven persons buried in timber on its bottom. It seems probable that the chief of an alliance of tribes and his favourites were buried in the grave. The construction of big barrows in Salbyk probably was based on the astronomical knowledge of the time. The installation of the fence slabs is connected with the main positions of the rising and setting of the moon and sun on astronomically significant days.

This is a preliminary version of the astronomical research that will be presented separately by the author and the astronomers V. L. Gorshkov and V. B. Kaptsjug of the Pulkovo Observatory, St. Petersburg.

Introduction
The Great Salbyk barrow is the best known of the megalithic monuments in Siberia (Fig. 1). The barrow is situated 65 km northward of the town of Abakan in Khakasia (Russia, Siberia). There are more than 50 big and middle-sized barrows, as well as many small ones. Salbyk, meaning puddle or rain-water in the Khakasian language, is situated in a natural valley limited on three sides by rather small steep slopes and on the northern side by the spurs of the mountain range of Kuznetsky Alatau. The archaeologist S. V. Kiselev excavated the Salbyk barrow in 1954-56 (Kiselev 1956); it is the largest barrow in the valley as well as in Khakasia.

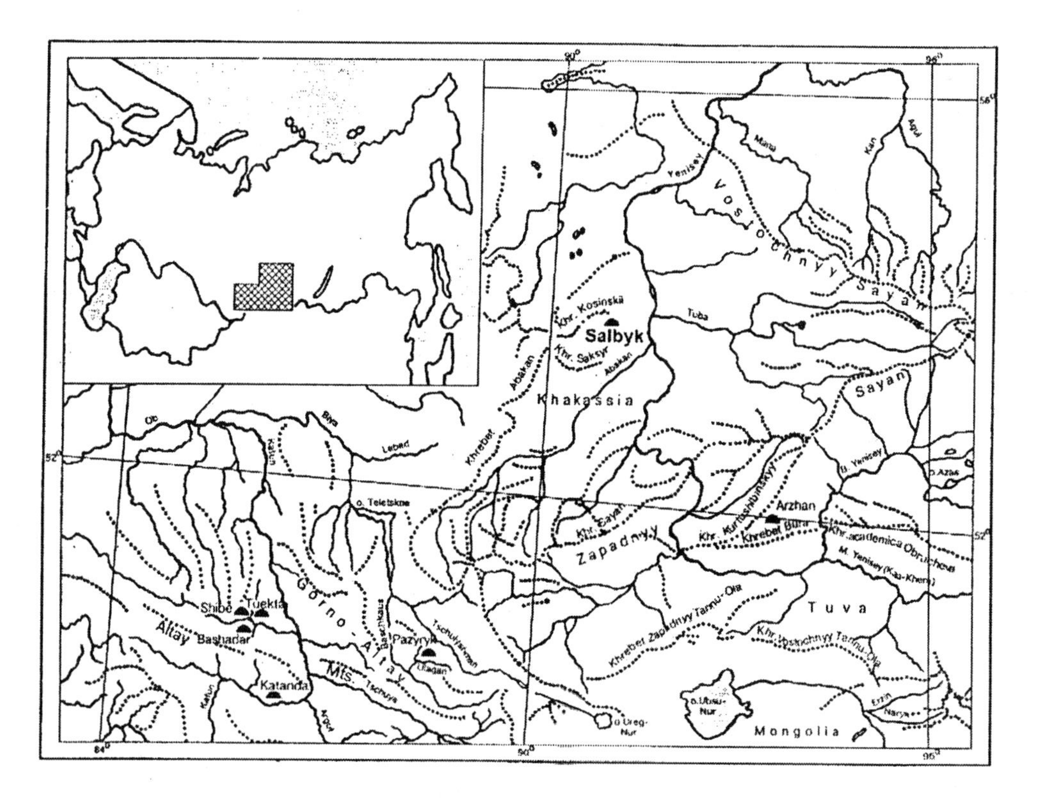

Fig. 1. Map of Saian-Altai.

The expedition of the State Hermitage Museum investigated the Salbyk valley in 1992, 1994, and 1996 (Marsadolov 1993; 2000). The expedition has recently composed a plan of the fence of stone slabs, taken samples for the tree-ring and radiocarbon analysis, conducted the astronomical and topographical analyses, and traced the connection of the barrow with the surrounding landscape. On the basis of the new analyses the barrow is dated to the 5th century BC.

Construction of the barrow

The process of building the great barrow in Salbyk, which is very complex due to its construction and the burial ritual, can be reconstructed in the following way. Probably not a whole year was spent in the building of it. Hundreds of persons took part in the construction of this monument-temple.

In the mountain valley a point for the barrow centre was chosen very carefully. The point had to satisfy certain requirements:

1) It had to be on the highest place between mountains.
2) It had to be surrounded by mountains lower to the west and east than to the north and south.
3) It had to be conveniently situated for astronomical observations.

It should be mentioned that such a place had probably been found in an earlier period and kept its meaning in later epochs. Near the Great Salbyk barrow there is the Bronze Age ritual centre with stone sculpture of the Okunevskaya culture, sites of the Early Tagar period, etc. After construction of the Salbyk barrow the cemetery near it of the Tashtyk culture continued to function for a long time.

The burial construction consisted of three parts: a corridor (dromos), an extension near a door, and a burial chamber. In the central part of the barrow, but closer to the western wall of the fence, a structure consisting of earth and logs was found. It looked like a reduced pyramid; its height was about 2 m and its upper platform was 8 x 8 m (Kiselev 1956; Devlet 1976; Kyzlasov 1985). The pyramid was at first sight snow white because its slopes were covered with a thick layer of birch bark (sometimes 15 layers). In addition the upper logs of larch were rolled up in birch bark. Thus larches were given the appearance of birches.

Under the pyramid there was a pit 5 x 5 m square and its bottom was 1.8 m deep. Its walls were lined with vertical logs. In the bottom of the pit there was a wooden framework with 4 rows of larch logs cut like bars. The chamber was 4 x 4 m; its height was about 2 m. It was covered by six rows of massive logs with a thick layer of birch bark. The bottom of the framework and the space between its walls and the logs covering the pit's walls were full of solid red water-resistant clay. Kiselev (1956) established that the bottom was covered with birch bark under the clay. Six layers of crossed logs formed a roof under the framework, but they could not withstand the great pressure of the earth and had fallen into the chamber.

In the chamber the remains of seven persons were found, men and women. An old warrior was buried in the centre, some of whose bones were broken. A large clay vessel was found in fragments. Near the middle part of the western wall of the framework, on the bottom, a miniature bronze knife was also found.

The corridor was a very high construction without a covering. The dromos began with an entrance near the middle stele of the western wall of the fence and went close to the western slope of the pyramid. There was a narrow hole into the chamber, which appeared to be filled with pieces of wood. The walls of the corridor (width 2-3 m) were covered with logs, and the upper parts of these were covered with a thin ceiling made of hewn planks. Both walls and the corridor's ceiling were decorated with a layer of birch bark.

The diameter of the barrow's circle in its lower part was about 500 m; its height was 11 m. The wooden fence was built first, and then the square fence measuring 70 x 70 m was constructed; it was oriented NEE, with a small deviation (Fig. 2). Astronomical observations had probably taken place there before the construction of the stone fence. The largest vertical stelai had been placed at the points connected with the astronomically important days. They were oriented by their angles to moonrise and moonset, and the entrance into the barrow was oriented to sunrise (Fig. 3).

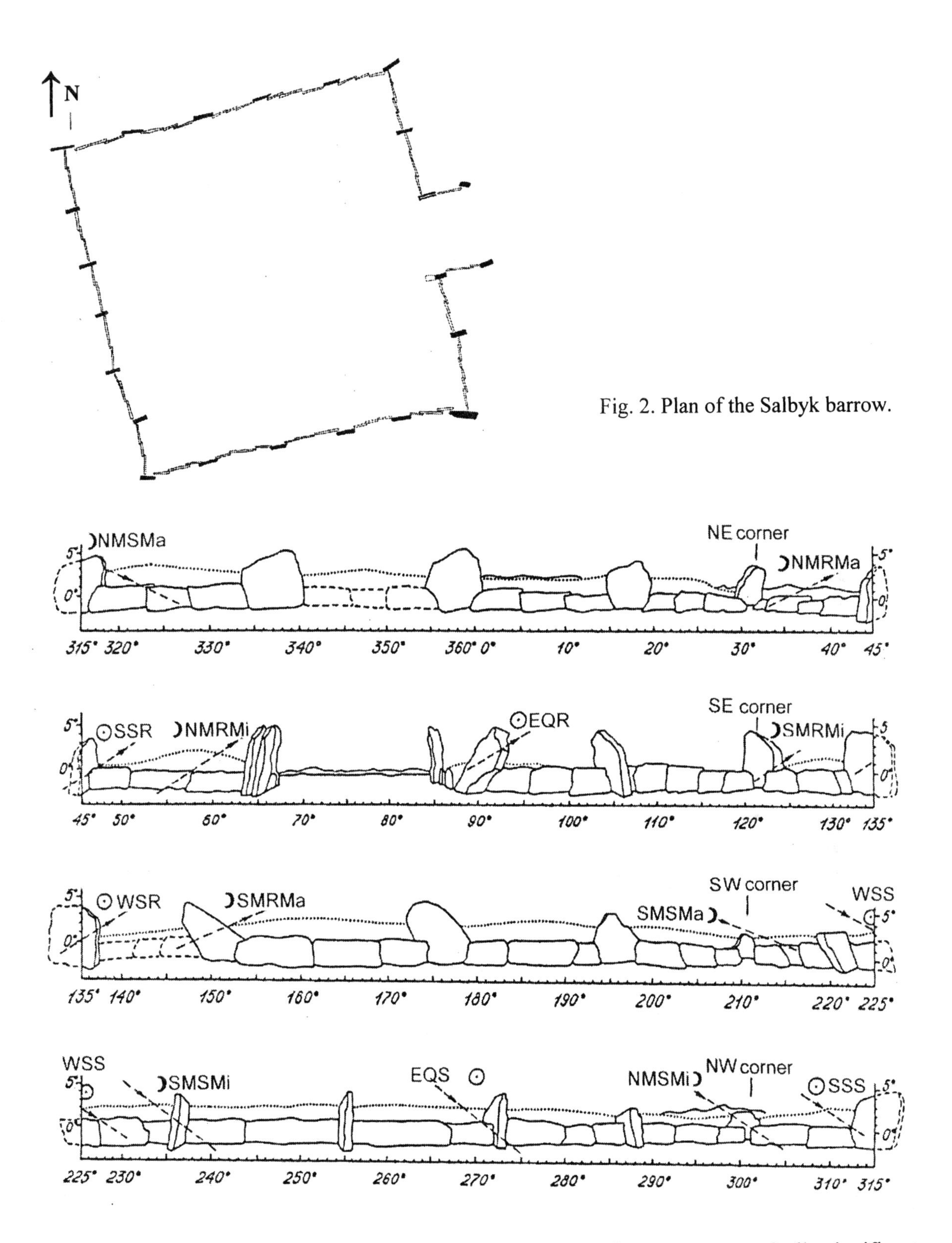

Fig. 2. Plan of the Salbyk barrow.

Fig. 3. The Salbyk barrow. The rising and setting of the moon and sun on astronomically significant days. Abbreviations:

Sun: SSR—summer solstice sunrise; SSS—summer solstice sunset; WSR—winter solstice sunrise; WSS—winter solstice sunset; EQR—equinox sunrise; EQS—equinox sunset.

Moon: MS—moonset; MR—moonrise; Ma—major standstill; Mi—minor standstill; NMRMa—northern moonrise, major standstill; NMRMi—northern moonrise, minor standstill; NMSMa—northern moonset, major standstill; NMSMi—northern moonset, minor standstill; SMRMa—southern moonrise, major standstill; SMRMi—southern moonrise, minor standstill.

Fig. 4. The Salbyk barrow: 1) the stone 'gate'; 2) corner of the stone slabs.

The fence was made of massive blocks of sandstone placed in a standing position (Fig. 4); the largest of these weigh about 30 tonnes. In the middle part of the eastern stele, there were two walls of 5 m to the east; they seemed to mark an entrance. The fence's blocks were put in narrow deep trenches, their width being less than 0.60 m. The depth of the trenches varied—from 0.8 to 2 m, depending on the height of the blocks, which were placed in such a way that they were about even above ground, with heights of 1.8 to 2 m. The entrance from the barrow's eastern side was rather complex. From the two middle stelai the long slabs were perpendicular to the line of the wall, resting on two stelai placed towards the east. From the eastern side 'the entrance' between the stelai was covered with a roof of small slabs placed very carefully, with a small deviation inside the barrow (Fig. 4).

During the clearing of the western wall the remnants of the much-destroyed burials of two persons were found in the SW and NE angles. In the process of studying the fence near the large angle of the southeastern stele, the destroyed burial of a child was found. In the angle formed by a block of the southern fence and the angle of the SW stele, the burial of an adult man with tied bent legs was discovered. These were probably sacrifices that had been made in the most significant places of the barrow.

The primary shape of the barrow was pyramidal. The mound formed a square 70 x 70 m inside the fence; its height was 25-30 m. The construction of the mound was gradual. First of all the greatest part of the mound was made of layers of turf. In the northern, western and southwestern parts, the mound was covered in some places with light-brown earth.

The transportation of blocks for the fence must have been very hard work. The barrow was erected in the steppe valley and the nearest deposits of Devonian sandstone are situated in the Kyzyl-Khaya mountain, 20 km south-westward from it. The remnants of ancient quarries were found there. The blocks were quarried from the rock, probably with the assistance of wooden wedges, and were transported to the place of the barrow construction, perhaps in winter on wooden rollers. During the clearing of the barrow's lower part many larch logs, often dry from forest fires, were discovered. The logs had been moved by ropes, judging from marks on their ends.

Astronomical aspects

The installation of the fence slabs is connected with the main positions of the rising and setting of the moon and sun on astronomically significant days (Fig. 3). Signs in the form of circles, crescents, and other figures, situated with respect to significant astronomical directional lines, were discovered on the barrow's slabs. The positions of the sun and moon moving by 'arc' were shown. The investigation revealed that the solar directions were connected with vertical stone slabs, but those of the moon with both vertical and horizontal slabs. The astronomers V. L. Gorshkov and V. B. Kaptsiug participated in the work of the expedition.

Fig. 5. The stone slab with drawings from the Salbyk barrow.

On one of the slabs from the barrow a complicated composition is drawn (Fig. 5). In the higher part of the slab the sky is represented: a bird, the sun, stars, a person with vizier in hand. In the middle part of the slab a male warrior stands with a foot on the head of a fallen person, nearby is a moon-woman and also a man. In the lower part of the slab there are unclear figures of perhaps a horse and a beast. It is possible that on this slab the sequence of the funeral ritual is represented, which corresponds to the archaeological material from the excavations.

In Salbyk some of great barrows have 'chains' of vertically standing slabs as well as horizontally placed 'slab-altars' near the mound. Outside the barrow were found vertical stones of intermediate size, aligned to astronomically significant directions. A sculptural representation of a reclining tiger was also found. The detailed study of the stone slabs of the fence revealed the significance of a colour spectrum, from light to dark tones and conversely.

The 'chain' of barrows in the Salbyk valley is oriented on a line northwest to southeast, the line of the extreme positions for moonrise and moonset. The location of barrows in Salbyk is principally distinguished from the orientation of barrows behind the Saian range. Near the Arzhan settlement in Tuva the great barrows (6th-5th centuries BC) were erected on a line northeast to southwest and oriented to the sun—to the high point of sunrise on the day of the summer solstice and the low point of sunset on the day of the winter solstice. Thereby the orientation of the barrow's chains serves the important additional (religious) criterion for two earlier-chosen large areas of the archaeological sites.

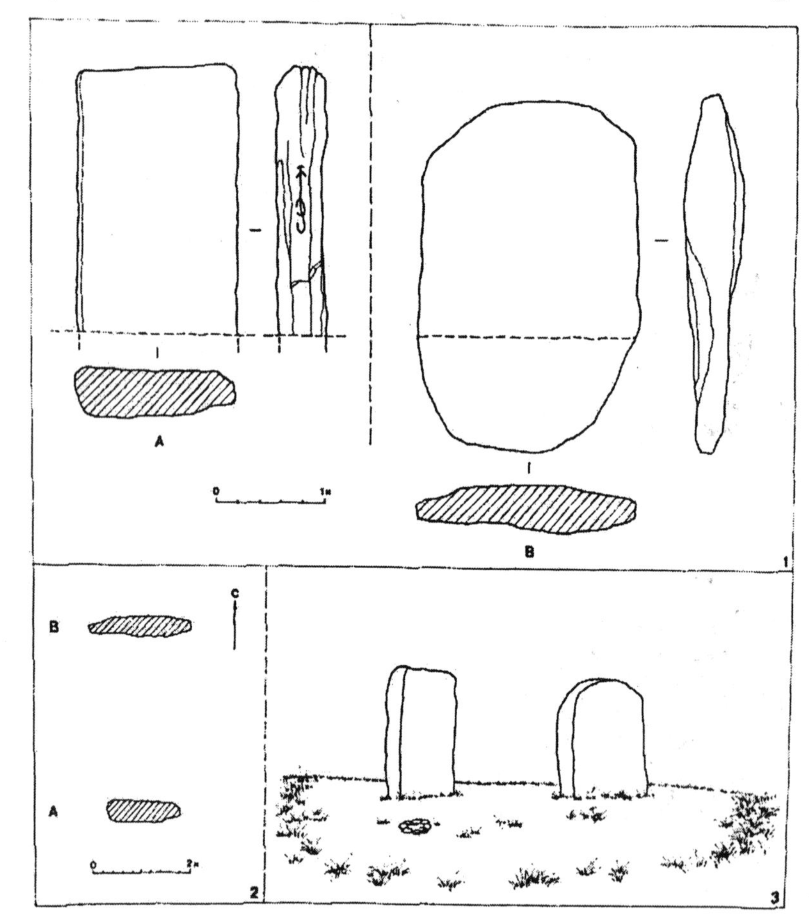

Fig. 6. The Salbyk's stone 'gate'.
1) plans and sections of the stone slabs.
2), 3) reconstruction of the situation of Salbyk's gate'.

Salbyk's 'gate'

Six km to the northeast of the Great Salbyk barrow, in the mountain valley, the original monument called Salbyk's 'gate' is situated (Fig. 6). Two vertical stone slabs were orientated with their narrow sides on an east-west line. The slabs have different colours, the southern slab is of grey siliceous sandstone and the northern one is of red-brown sandstone. The northern slab, now fallen, is oval in shape and the southern one is almost square-shaped, with a small slope in the higher part (with high eastern edge, lowering westward like the majority of the Tagar barrow's slabs). The narrow high eastern edge of the standing southern slab points out the border of 'slopes' of the high neighbouring southern mountainside and the low distant plain with the mountains on the horizon line. The low western edge of the slab is directed to the low, very distant part of the horizon, and the group of middle-sized barrows are situated 6.5 km westward from the 'gate'.

Conclusions

The construction of big barrows in Salbyk having multiple functions (funeral, socio-political, religious, astronomical, architectural, and others) probably was based on the astronomical knowledge of their time.

The Great Salbyk barrow, by its monumental construction, can be put in the same group as the famous Stonehenge in England; but by the volume of consumed labour, it probably significantly exceeds Stonehenge.

*The State Hermitage Museum, Saint Petersburg, Russia, marsadolov@hermitage.ru.

References

Devlet, M. A. 1976. The Great Salbyk barrow: A grave of a tribal chief (in Russian). In *From the History of Siberia*, 146-154. Tomsk.

Kiselev, S. V. 1956. The study of the Great Salbyk barrow in 1954 and 1955 (in Russian). In *Abstracts of reports of the session of the Department of Historical Sciences and plenum of the Institute of the History of Material Culture devoted to the results of archaeological studies in 1955*, 56-58. Moscow.

Kyzlasov, I. L. 1985. *The pyramids of the Salbyk valley* (guidebook in Russian). Abakan.

Marsadolov, L. S. 1993. The works of the Sayan-Altai archaeological expedition in 1992 (in Russian). In *The summary of the archaeological session of the State Hermitage Museum (abstracts)*, 3-5. St. Petersburg.

Marsadolov, L. S. 2000. *Archaeological sites of the 9th-3rd centuries BC in the mountain regions of Altai as cultural-historical sources (the phenomenon of the Pazyryk culture)* (in Russian). Thesis for the Ph D in Culturology. St. Petersburg.

The Minoan peak sanctuary on Pyrgos and its context

Mary Blomberg* and Göran Henriksson**

Abstract

We report the results of our archaeoastronomical study of the small Middle Minoan building on Pyrgos (Maleviziou), the mountain near the large Minoan villa at Tylissos. The orientation of the long wall of the building indicates that the major axis was aligned to sunrise at the summer solstice. This is the second example of such an orientation for a peak sanctuary in Crete, the other being that of the major axis of the structure on Petsophas. The short axis of the building on Pyrgos is oriented to where the heliacal setting of Arcturus occurred at the end of the Early Minoan Period. Arcturus is one of the four brightest stars, and we know from surviving texts that it was an important calendar star in the Aegean from very early times. This is the third example of orientations to Arcturus at the Minoan peak sanctuaries, the other two being at Petsophas and Traostalos. At the beginning of the Middle Minoan Period (ca 2000 BC), the heliacal setting of this star as seen from Pyrgos would have occurred directly above the prominent peak of Kako Kefali, which thus could have served as a foresight in the same way that the peak of Modi could have for the heliacal setting of the same star as seen from Traostalos.

Conclusions based on the accumulating evidence from orientations of the Minoan palaces, villas, and peak sanctuaries included in the Uppsala Project are summarized and compared briefly with the differences in grave orientations.

Methods

We apply archaeoastronomical methods to the study of the Minoan archaeological remains; this means measuring the orientations of the buildings and evaluating the data using appropriate tools. These are the basic methods of archaeoastronomy (Hoskin 2001: 7-20; Blomberg and Henriksson 2001b: 609-610) and, as far as we know, they are the only ways of recovering the astronomical achievements of an ancient culture from which no legible documents have survived. It is fortunate that peoples in many places and periods have been strongly motivated to establish physical relationships between themselves and the sky by orienting their settlements and buildings to prominent celestial events, thus leaving this evidence of their astronomical interests.

Crucial to the accuracy of the results obtained using archaeoastronomical methods is reliance on adequate measuring equipment, computer programs, and input parameters. For measuring orientations we use the digital theodolite SOKKIA SET 4C; for the computations of the astronomical data we use the computer programs developed by Henriksson. The parameters for calculating the visibility of bright stars are from Bemporad (1904), Sidentopf (1941), Ljunghall (1949), and Schmidt (1865). It is important to use Schmidt's visibility calibrations for Athens from ca 1850, as his observations were made before modern air pollution. When statistical evaluations are relevant, we rely upon the Department of Mathematical Statistics at Uppsala University (Henriksson and Blomberg 1996: 111; 2000: 307).

Pyrgos in the context of the Uppsala project

Our study of the peak sanctuary on Pyrgos is part of the on-going Uppsala project to collect the evidence for Minoan astronomical observations, to interpret this evidence, and to evaluate its influence on the Minoan culture and on the later Mycenaean and Greek cultures. We have measured the alignments of fifteen Minoan monuments (Blomberg and Henriksson 2001a; 2002; Henriksson and Blomberg 1996; 1997-1998; forthcoming). These include four palaces, five villas, and six peak sanctuaries (Fig. 1). We have not yet completed our evaluation of all sites so that the results presented here are preliminary.

There may have been as many as 25 Minoan peak sanctuaries, but it seems that fewer than ten ever had buildings. The identification of those with no buildings has been made on the basis of the types of small objects found on the sites (Rutkowski 1986: 73-98). The construction of these places probably dates to early in the Middle Minoan Period, which began ca 2000 years before our era. It seemed to us that sites of this type, especially those on the eastern coast of Crete, were ideally placed for the purpose of observing the celestial bodies, and we therefore studied the orientations of their walls and examined the finds to see if these could support such a function. Our study of the peak sanctuary on Pyrgos is the fourth that we have completed, the other three begin on Juktas, Petsophas, and Traostalos (Blomberg and Henriksson 2002; Henriksson and Blomberg 1996; 1997-1998).

Fig. 1. Minoan sites in the Uppsala University archaeoastronomical project.

The archaeology of Pyrgos

The site of Pyrgos with its small building was excavated in 1962 by the local superintendent of antiquities for the Greek Archaeological Service. It is located in north central Crete on a moderately high mountain (684 m), and this is typical of such places; they are not on the higher mountains of Crete (Rutkowski 1986: 73-74; Peatfield 1990: 119-120). The finds consisted of pottery shards, human and animal figurines, a layer of ashes, and part of a pair of horns (Alexiou 1963; Megaw 1962/1963: 31). In 1997, wall BE of the building was 12.8 meters long, and wall CD measured 5.6 meters (Fig. 2).

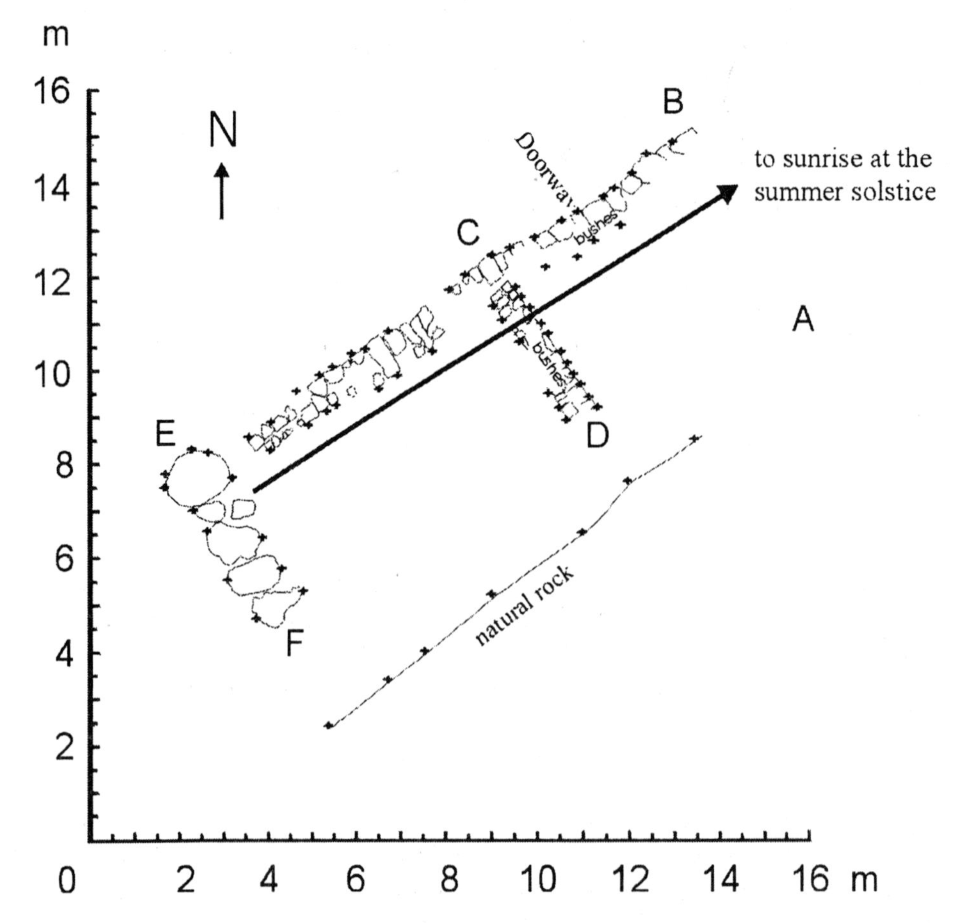

Fig. 2. Plan of the peak sanctuary on Pyrgos, north central Crete. The orientation of the wall BCE is 59.3° ± 0.4° and that of the wall CD is 325.2° ± 0.2°.

In the summer of 2000, they were 11.2 and 3.2 meters respectively. This destruction is due to the illegal digging for valuable objects and, unfortunately, is the fate of many Minoan monuments in remote areas, making the documentation of these places urgent. We do not know if there had been a built rear wall or if the natural rock screen served that purpose. The rock seems to have been somewhat smoothed with tools. The peak sanctuary on Traostalos on the eastern coast of Crete clearly had smoothed natural rock as one of its walls. It has been suggested that these structures were, at least in part, open to the air, but we have no proof one way or the other.

The archaeoastronomy of Pyrgos

Using orthogonal regression estimation, we calculated the orientation of the long wall BCE and found it be 59.3° ± 0.4° (Fig. 2). The calculated azimuth of the upper limb of the sun at sunrise at the summer solstice in 2000 BC is 59.2° (Fig. 3). The horizon is the open sea, and the altitude at which the sun will become visible is sensitive to variations in atmospheric refraction and extinction. From Pyrgos sunrise at the summer solstice in the Middle Bronze Age could have been observed as far north as we see it in Fig. 3, but never further north. The computed horizon altitude is -0.39° from a mountain with height of 684 m (that of Pyrgos). The refraction is valid for a temperature of +17°C and barometric pressure of 760 mm of Mercury.

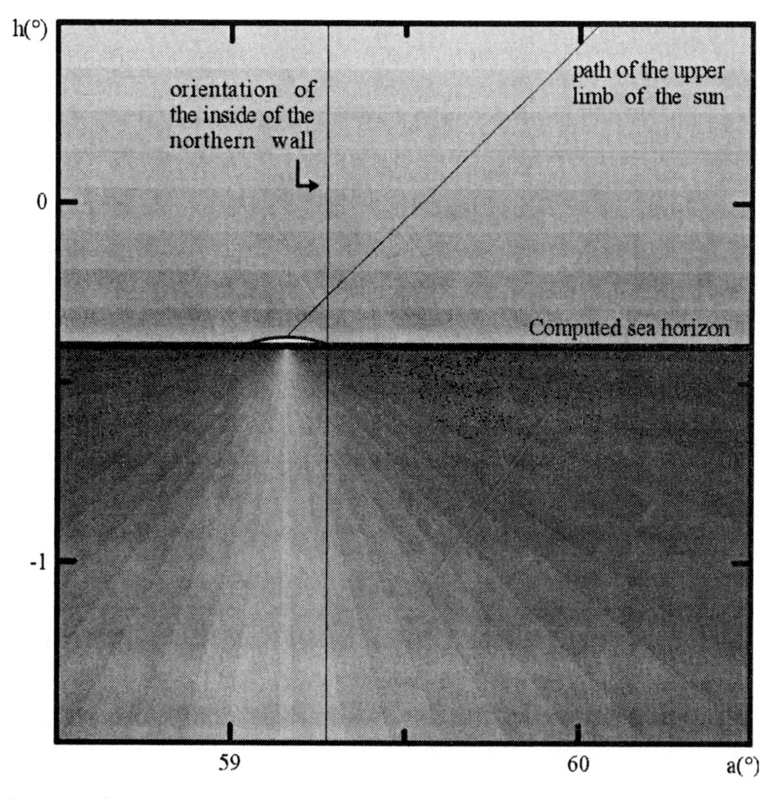

Fig. 3. Sunrise at the summer solstice on 23 June 2000 BC, 04.34.38 local mean solar time, from the peak sanctuary on Pyrgos.

The weighted mean value of the minor axis of the building is 325.2° ± 0.2°. This is the orientation to the heliacal setting of Arcturus as it would have been observed from Pyrgos towards the end of the Early Minoan Period (Fig. 4). The last phase of the period began about 2200 BC. At that time the heliacal risings and settings of Arcturus were very close to the limits of the circumpolar stars for the latitude of Pyrgos (35°19'). These stars are important for navigation and they, as well as Arcturus, have figured prominently in early Greek literature, especially in connection with navigation. The hypothesis has been presented that Calypso's reference to the Bear in book 5 of the *Odyssey* is actually a reference to the circumpolar stars as a whole (P. Blomberg this volume), the word for bear in Greek meaning also north (ἄρκτος, compare our 'arctic').

To someone standing near the building on Pyrgos, the most prominent peak opposite the site, that of Kako Kefali, would have been a foresight for the heliacal setting of Arcturus at the beginning of the Middle Minoan Period, ca 2000 BC, when the structure was most probably built.

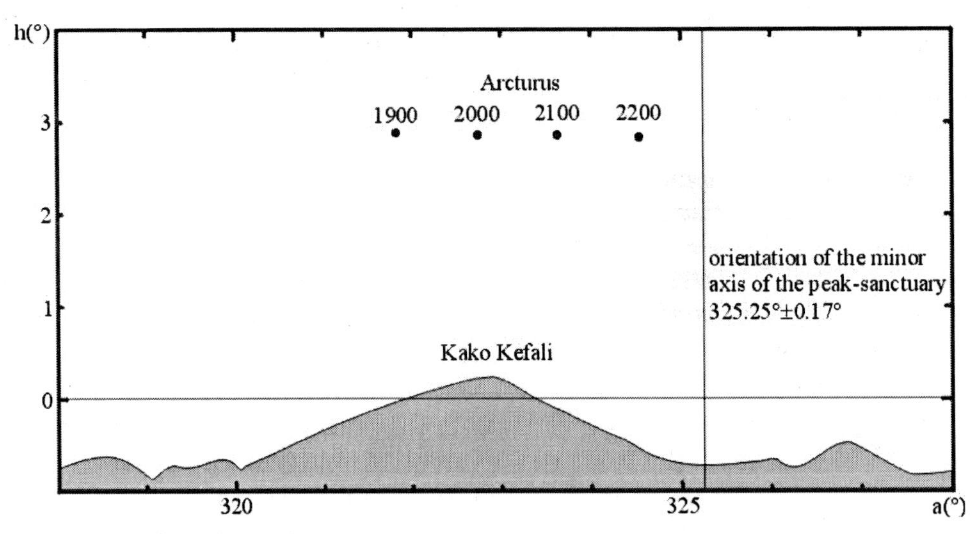

Fig. 4. The heliacal settings of Arcturus during the late Early Minoan Period and the early Middle Minoan Period. The changes in position are due to the phenomenon of precession.

Pyrgos compared with Petsophas and Traostalos

The orientations measured on Pyrgos gain significance when they are compared to those that we found earlier at the peak sanctuaries on Petsophas and Traostalos on the eastern coast of Crete. The long wall of the building on Petsophas was oriented, as at Pyrgos, to sunrise at the summer solstice (Fig. 5), and the highest peak on the island of Karpathos opposite would have been an excellent foresight for this event (Fig. 6). There are also two oblique walls at the site, AA' and AB, that were oriented to the heliacal rising and setting of Arcturus in the same period. The horizon is the open sea in both directions (Henriksson and Blomberg 1996).

At Traostalos (Fig. 7) there are also two oblique walls (AA' and AB) that were oriented to the horizon setting and rising respectively of Arcturus in the same period and there is another wall (BC) oriented to the eastern limit of the circumpolar stars. Traostalos lies about 7 km due south of Petsophas, and the entire eastern coast of Crete can be seen from it in clear weather. The orientation on Traostalos to the setting of Arcturus could have had as foresight the isolated conical peak of Modi. Thus we have a third instance of orientation to Arcturus—the other two being at Pyrgos and Petsophas—and a second instance of a prominent mountain peak that would have made a very good foresight for the heliacal setting of the same star (Henriksson and Blomberg 1997-1998).

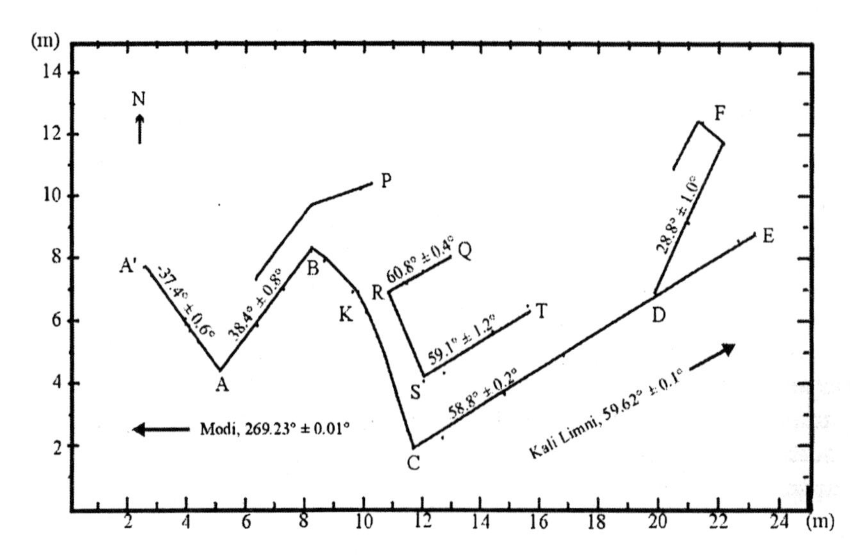

Fig. 5. Plan of the peak sanctuary on Petsophas showing orientations to the heliacal rising (AB) and setting (AA') of Arcturus and to sunrise at the summer solstice (CDE). From the near vicinity of the building sunset at the equinoxes could have been observed directly behind the isolated conical mountain peak of Modi.

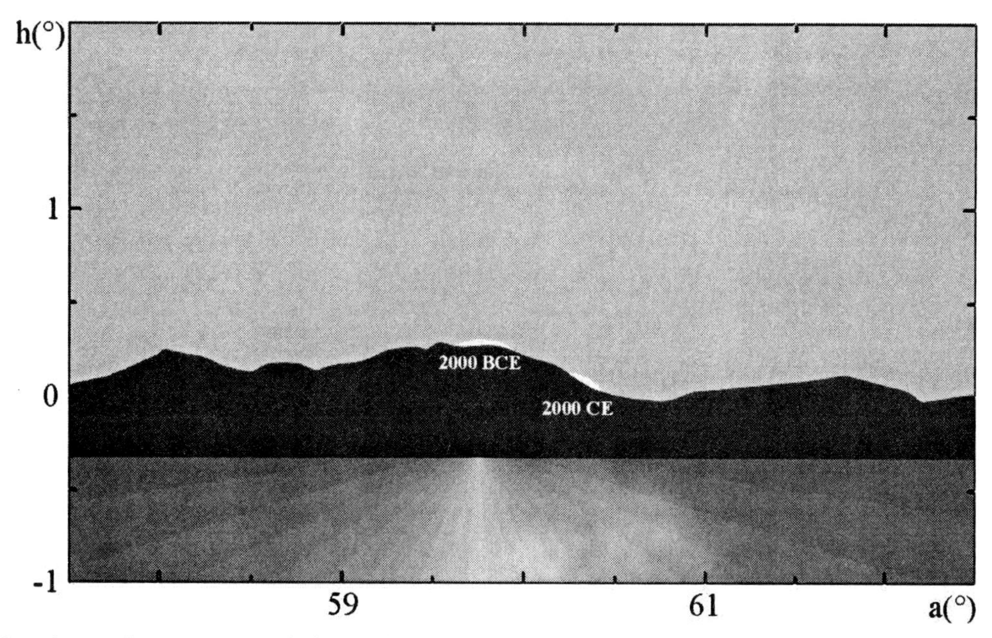

Fig. 6. Sunrise at the summer solstice as it would have been observed along the main axis of the peak sanctuary on Petsophas in the years around 2000 BC. The highest peak of the mountain Kali Limni on the island of Karpathos could have served as a foresight.

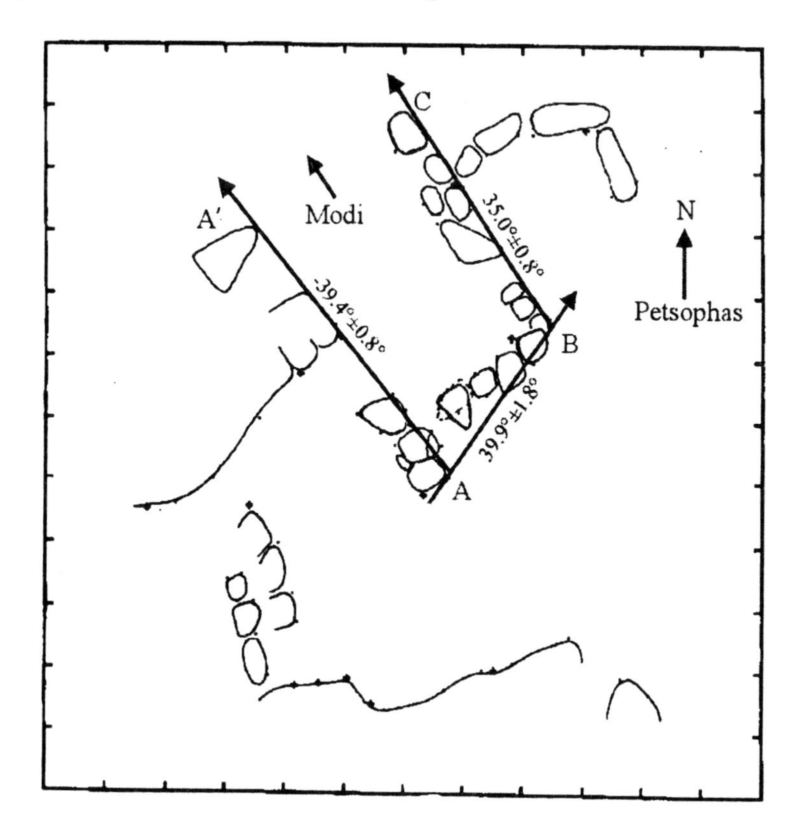

Fig. 7. Plan of the peak sanctuary on Traostalos showing orientations to the heliacal rising (AB) and setting (AA') of Arcturus and to the eastern limit of the circumpolar stars (BC).

Summary of the orientations of sites included in the Uppsala project

We have summarized in Fig. 8 the orientations and other relationships that we have found so far in our study of fifteen major Minoan monuments: the four major palaces, the six peak sanctuaries with adequate surviving walls, and five of the large villas. The term 'relationships' is used because, in a few cases, there is not an orientation of an architectural feature; but the observation of the celestial event would only have occurred in the near vicinity of the building, e.g. sunset at the equinoxes behind Modi as observed from Petsophas (Henriksson and Blomberg 1996: fig. 2). Also, we must keep in mind that the azimuths of the orientations are in most cases influenced by the mountainous landscape, being shifted southwards due to the altitude of the horizon.

Our study has given striking results. Ten of these fifteen monuments have orientations to major celestial events: sunrise at the summer solstice in two cases, sunrise at the equinoxes in four cases, sunset at the equinoxes in one case, sunset at the summer solstice in three cases, sunrise at the winter solstice in one case, moonrise at the southern major standstill in one case, the heliacal rising of Arcturus in two cases, and the heliacal setting of the same star in three cases, a total of seventeen orientations to major celestial events at these ten sites. It is a striking fact that in the case of eleven of these seventeen orientations there were features that would have made excellent foresights; seven were natural features and four were man-made. The asterisks in Fig. 8 signify the presence of such a feature. Four of the monuments have orientations to more than one major celestial event: two each in the case of the peak sanctuaries on Pyrgos and Traostalos, three in the case of the villa at Vathypetro, and four in the case of the peak sanctuary on Petsophas. Sunrise at the solstices and equinoxes are all objects of Minoan orientations. Sunset at these times seems not to have been as important and those three to sunset at the summer solstice—the oblique building at Mallia, the tripartite shrine at Vathypetro, and the small shrine at Agia Triada—are therefore especially interesting and have been discussed in separate publications (Blomberg and Henriksson 2001a; forthcoming). The peak sanctuaries show a greater number of different relationships and only they have orientations to the important calendar star Arcturus. In the Middle Bronze Age it rose and set at the times later known to have been the limits of the sailing season in the Aegean. Furthermore there are orientations at two peak sanctuaries, Pyrgos and Traostalos, which mark the limits of the circumpolar stars for their latitude. The orientation at Petsophas to sunset at the equinoxes is also unique and is due, we think, to the special function of that site for observing the celestial bodies—for the sake of keeping the calendar, for navigation, and perhaps also for religious reasons (Henriksson and Blomberg 1996: 113). Another unique orientation is that of the palace at Zakros, which is oriented to moonrise and with a natural foresight that could have marked the alignment. The villas are oriented to the east within the limits of sunrise, but we have not completed our study of these places and, aside from Vathypetro, we do not yet know their precise relationships to sunrise.

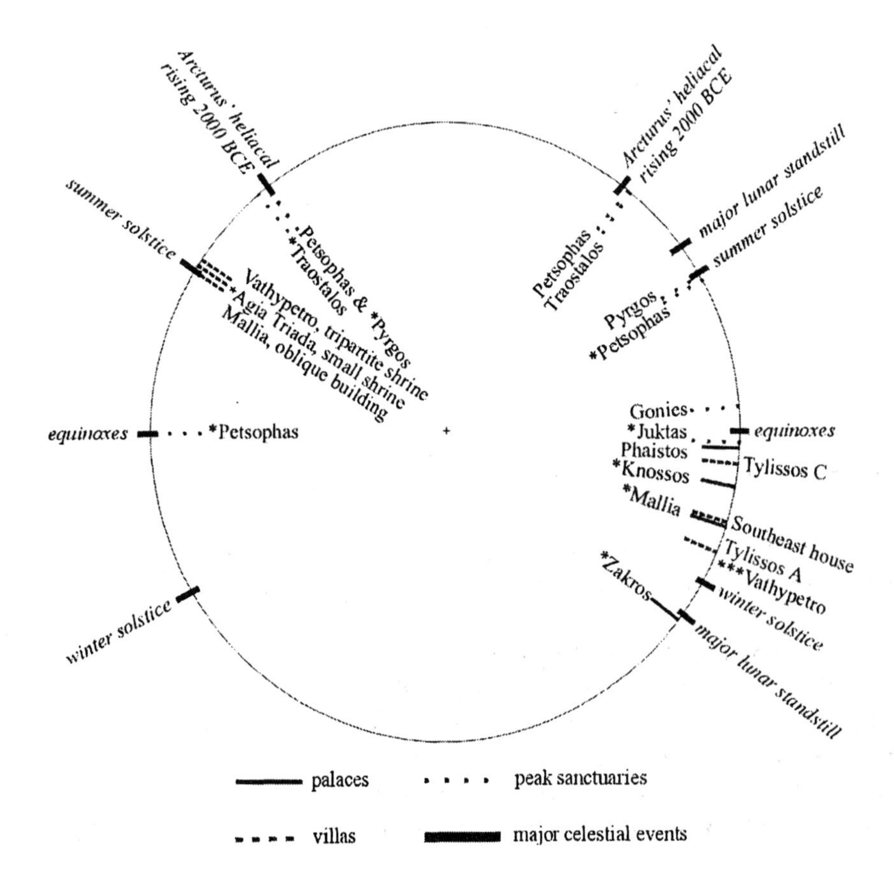

Fig. 8. The Orientations of 15 major Minoan monuments: 4 palaces, 6 peak sanctuaries, and 5 villas.

Grave orientations in Minoan Crete

In our study of 323 graves with passage approaches at 15 sites in Crete (Blomberg and Henriksson 2001a: 77-84), we found an overwhelming number with orientations to the east within the limits of sunrise (Fig. 9): 86% lie within these limits, which is only 18% of the circle, and 40% lie within ± 10° of due east, which is less than 6% of the circle. From these data it is not possible to tell whether the orientations were determined by religious or other factors, but the insistence on orientation to within the limits of sunrise, or possibly moonrise, is clear.

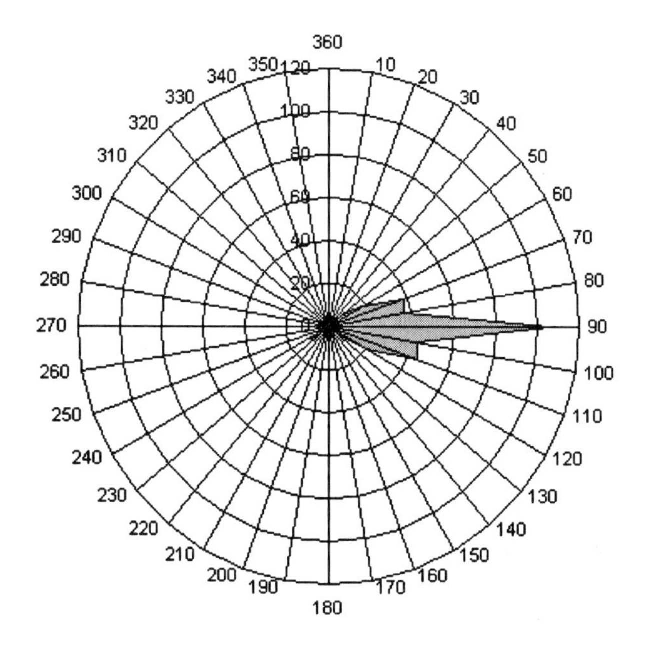

Fig. 9. Frequency distribution of the orientations of 323 chamber tombs with passages from 15 sites in Crete.

Fig. 10. Female figurine from Petsophas. Archaeological Museum, Heraklion, Inv. no. 3431. Photograph and permission to publish courtesy of the museum.

Conclusions

This fact of the differences in orientations for different types of structures in Minoan Crete should warn us against offering a single interpretation for their meaning, whether religious or practical. We are inclined to conclude from our results that Pyrgos and the other Minoan peak sanctuaries, especially those on the eastern coast, had as an essential part of their function the systematic study of the motions of the celestial bodies and that this was primarily for practical purposes: for regulating the calendar and for navigation. The figurines found at these sites support such a function, as Peter Blomberg (2000) has shown. The orientations of the four palaces, on the other hand, suggest a religious significance connected with celebrations of important events in the year, e.g. the new year and the harvest. A religious meaning for the orientation to the moon at Zakros is supported by the large number of figurines of women with moon-shaped heads (Fig. 10) found at the nearby peak sanctuaries on Petsophas and Traostalos and elsewhere in the eastern part of the island. The strong concentration to the east for the orientations of graves in Minoan Crete seems to us to have been motivated also by religious beliefs, most likely centering on ideas concerning the regenerative power of the sun.

*Norrtullsgatan 31, SE-113 27 Stockholm, Sweden, mary@mikrob.com.
**Astronomical Observatory, Uppsala University, Box 515, SE-751 20 Uppsala, Sweden, goran.Henriksson@astro.uu.se.

References

Alexiou, S. 1963. Χρονικά. *Κρητικὰ χρονικά* **17**, 404-405.

Bemporad, A. 1904. Zur Theorie der Extinktion des Lichtes in der Erdatmosphäre. *Mitteilungen Grossh. Sternwarte zu Heidelberg* **4**, 1-78.

Blomberg, M. and Henriksson, G. Forthcoming. Literary and archaeoastronomical evidence for the origins of the Hellenic calendar in the Aegean Bronze Age. Paper read at the 8th annual meeting of the European Association of Archaeologists, 24-29 September 2002, Thessaloniki.

Blomberg, M. and Henriksson, G. 2001a. Differences in Minoan and Mycenaean orientations in Crete. In *Astronomy, Cosmology and Landscape*, proceedings of the SEAC 98 meeting, Dublin, Ireland, September 1998, ed. C. Ruggles, F. Prendergast, and T. Ray, 72-91. Bognor Regis.

Blomberg, M. and Henriksson, G. 2001b. Archaeoastronomy: new trends in the field, with methods and results from studies in Minoan Crete. *Journal of Radioanalytical and Nuclear Chemistry* **247**, 609-619.

Blomberg, M. and Henriksson, G. 2002. The calendaric relationship between the Minoan peak sanctuary on Juktas and the palace at Knossos. In *Proceedings of the conference "Astronomy of Ancient Civilizations" of the European Society for Astronomy in Culture (SEAC) and National Astronomical Meeting (JENAM), Moscow, May 23-27, 2000*, ed. T. M. Potyomkina and V. N. Obridko, 81-92. Moscow.

Blomberg, P. 2000. An astronomical interpretation of finds from Minoan Crete. In *Oxford VI and SEAC 99: Astronomy and cultural diversity*, ed. C. Esteban and J. A. Belmonte, 311-318. Santa Cruz de Teneriffe.

Blomberg, P. 2003. The northernmost constellations in early Greek tradition. This volume.

Henriksson, G. and Blomberg, M. 2000. New arguments for the Minoan origin of the stellar positions in Aratos' *Phainomena*. In *Oxford VI and SEAC 99: Astronomy and cultural diversity*, ed. C. Esteban and J. A. Belmonte, 303-318. Santa Cruz de Teneriffe.

Henriksson, G. and Blomberg, M. 1996. Evidence for Minoan astronomical observations from the peak sanctuaries on Petsophas and Traostalos. *Opuscula Atheniensia* **21**, 99-114.

Henriksson, G. and Blomberg, M. 1997-1998. Petsophas and the summer solstice. *Opuscula Atheniensia* **22-23**, 147-151.

Henriksson, G. and Blomberg, M. 2002. Some problems in Minoan archaeoastronomy. To appear in *Cultural context from the archaeoastronomical data and the echoes of cosmic catastrophic events, SEAC 2002, Tartu, 27-30 August 2002*, forthcoming.

Hoskin, M. 2001: *Tombs, temples and their orientations: A new perspective on Mediterranean prehistory.* Bognor Regis.

Ljunghall, A. 1949. The intensity of twilight and its connection with the density of the atmosphere. *Meddelanden från Lunds astronomiska observatorium*, ser. **2**, vol. **13**, no. **125**, 1949.

Megaw, A. H. S. 1962/1963. Tylissos. *Archaeological Reports* 1962/1963, 30-31.

Peatfield, A. 1990. Minoan peak sanctuaries: History and society", *Opuscula Atheniensia* **18**, 117-131.

Rutkowski, B. 1986. *The cult places of the Aegean.* New Haven and London.

Schmidt, J. F. 1865. Über die Dämmering. *Astronomische Nachrichten* **63**, article no. 1495.

Siedentopf, H. 1941. Neue Messungen der visuellen Kontrastschwelle. *Astronomische Nachrichten* **271**, 193-203.

Temples and astronomy in Carthage

César Esteban*

Abstract
We have found that the main axis of the urban layout of ancient Carthage is directed to the winter solstice sunrise. Similar orientations are found in religious buildings of other Punic cities, suggesting a possible astronomical orientation pattern. Astronomical motivations seem also to underlie the orientations of tombs of Punic necropolises and of temples dedicated to Saturn, the supreme god of almost all North African Roman territories and the inheritor of the ancient cult to the Punic god Baal-Hammon. We have explored the features of the almost total eclipse observed at Carthage on April 30, 463 BC, which has been conjectured to have had a special significance for the religious and even political evolution of the Punic state. Finally, we have found that the relevant total solar eclipse of August 15, 310 BC, was also coincident with a period of important religious changes in Carthage.

Introduction

The Phoenicians reached the shores of present-day Tunisia at the beginning of the first millennium BC, founding a chain of colonies and outposts from the Great Syrtes to the Atlantic coast. Among them Carthage was undoubtedly the most important and the capital city of Punic culture. The Punic people gained hegemony amongst the former Phoenician colonies of the western Mediterranean by the sixth century BC and extended their domination to the hinterland of Carthage in the following centuries. The impact of Punic culture on the North African indigenous substrate (the so-called Libyans or Proto-Berbers) was profound in many respects, especially on their religion. The cult to the most important god of the Punic pantheon, Baal-Hammon, was widespread in all North Africa and was probably the adaptation of a cult to a great ancient Libyan god.

We have very few data about the development of astronomy in Punic culture and most of this little knowledge comes indirectly from Phoenician and Canaanite sources. Stieglitz (2000) has proposed that the Phoenician-Punic calendar had twelve months with a periodic intercalation of a thirteenth month in the lunar year. This author also suggests that the year began in the autumn, the first month corresponding approximately to our October.

There are no extensive archaeoastronomical studies dedicated to Punic culture. The orientations of tombs in some Punic necropolises have been reported and analysed by Belmonte et al. (1998) and Ventura (2000). The orientations of Punic religious buildings have not been treated in any detail in the literature. This brief work is only a first attempt to accomplish such a study.

The Orientations of Punic temples

From the published maps of ancient Carthage (Hurst 1999; Lancel 1994), it is evident that the town was arranged in a quite regular pattern (Fig. 1). The *decumanus maximus* of Roman Carthage is oriented to the east, to an azimuth of about 120°, pointing to the local winter solstice sunrise. However this suggestive orientation could be simply due to the fact that it is perpendicular to the general orientation of the local coastline. It is important to note that the layout of the Roman city coincides with that of the previous Punic city, at least from the slopes of the Byrsa hill to the seashore (Lancel 1994: 131-146; Hurst 1999: 10). This fact indicates that the general orientation of the city was a Punic design. The Roman acropolis on Byrsa, the heart of Carthage, shows the same orientation. My colleague J. A. Belmonte and I measured the orientation of the foundations of the Roman acropolis in 1998, finding that it is consistent with the solstitial orientation. It is known that the Punic citadel on the summit of the Byrsa hill was physically cut away by the Romans, but the existence of a clear orientation in both the Roman and the Punic cities suggests that the orientation of the temple of Eshmun, on the top of Byrsa, was perhaps also oriented towards the winter solstice sunrise. Niemeyer (2000) has recently reported the finding of a Punic sanctuary dedicated to Tanit at the *decumanus maximus* of Carthage; this building has its entrance facing east and follows the general orientation of the city.

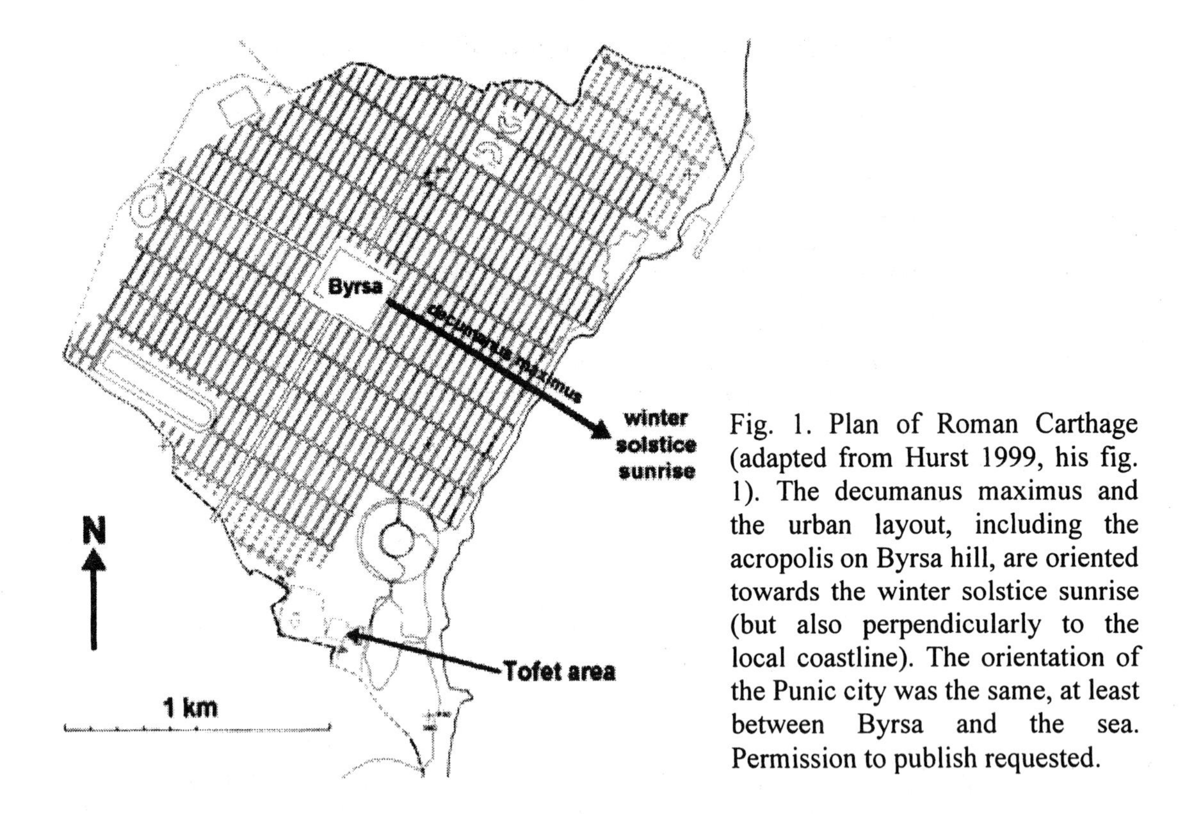

Fig. 1. Plan of Roman Carthage (adapted from Hurst 1999, his fig. 1). The decumanus maximus and the urban layout, including the acropolis on Byrsa hill, are oriented towards the winter solstice sunrise (but also perpendicularly to the local coastline). The orientation of the Punic city was the same, at least between Byrsa and the sea. Permission to publish requested.

We have made a study of the orientations of Punic sacred buildings in order to ascertain if the orientation of the layout of Carthage and the religious acropolis of Byrsa is, or is not, an isolated case. We have compiled published plans of several Punic temples and sanctuaries. The sample is very limited because of the few surviving remains of buildings known and published. As is rather usual, it is not clear whether the north indicated in some plans corresponds to the magnetic or to the geographic north. Therefore, in those cases, the orientations given are only indicative and accurate to within some few degrees. In Fig. 2 and Table 1 we show the orientations of the entrances of the Punic temples of Kerkouane—referred to the geographic north (Esteban et al. 2001), El Hofra—referred to the geographic north, which is indicated in the plan (Berthier and Charlier 1955), the Phoenician-Punic temple of 'Cappiddazzu' at Motya in Sicily (Tusa 2000), and the Neo-Punic temples at Thinnissut (Merlin 1910, plate 1) and El Kenissia (Picard 1954: fig. 17).

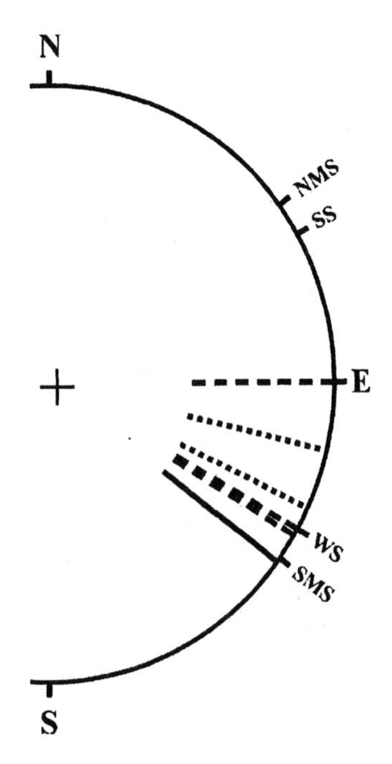

Fig. 2. Orientations of Punic temples and sanctuaries. Continuous line: direct measurement of the urban temple of Kerkouanne; dashed lines: orientations obtained from published plans; dotted lines: orientations of the external and internal parts of the sanctuary of El Kenissia, data obtained from a published plan. See text for references. SS and WS mean summer and winter solstice; NMS and SMS mean northern and southern major standstills of the moon.

From Fig. 2 we can see that the five temples are oriented towards a fairly narrow zone of the horizon, from ≈ 90° to 127°. It is remarkable that two of them, El Hofra and Motya, are oriented towards the winter solstice sunrise and that the temple of Kerkouane is oriented only a few degrees to the south of that event. The other two temples, Thinnissut and El Kenissia, show a somewhat more easterly orientation; these are the latest buildings, built in Roman times (first to second centuries AD), but dedicated to Punic deities (Picard 1954: 153-157). The fact that the three genuine Punic religious buildings of the sample are oriented about the winter solstice sunrise suggests a possible ritual significance of this particular direction and makes sense of the possibility of an astronomical motivation in the orientation of ancient Carthage.

Temple	Azimuth (°)	Declination (°)	Notes
Kerkouane	127 ± 2	−29 ± 2	measurements by Esteban et al. (2001)
El Hofra	~ 120	~ −24	from map (Berthier & Charlier 1955)
Motya	~ 119	~ −23	from map (Tusa 2000)
Thinnissut	~ 90	~ 0	from map (Merlin 1910)
El Kenissia	~ 115 ~ 103	~ −20 ~ −11	external part, from map (Picard 1954) internal part, from map (Picard 1954)

Table 1. Orientations of Punic temples.

There are other interesting facts about the orientations of Punic religious and funerary buildings. The general planning of the Punic sacred areas, or *tofets*, in Sicily and Sardinia is usually arranged along the cardinal axes, e.g. Selinunte, Solunto, Motya, Sulcis (Ribichini and Xella 1994). An easterly orientation is also reported in some Punic necropolises, as at Utica and the hypogea of Menzel Temine (Belmonte et al 1998). It is remarkable that a similar doubled-peaked distribution centred on sunrise at the equinoxes and the winter solstice has been found in Menzel Temine and the early Phoenician necropolis of Villaricos on the southern Mediterranean coast of the Iberian Peninsula (Belmonte 1999, fig. 5.5). A general east-west orientation is also found in the Punic eastern coastal necropolis of Tipasa in Algeria (Baradez 1969) and the Phoenician/Punic necropolis of Aïn Dalia Lekbira in northern Morocco (Alaoni 2000). However there is not a regular orientation in any of the Phoenician/Punic necropolises studied. In Byrsa (Carthage), Belmonte et al. (1998) have found a rather unusual south-west distribution. In the case of the large necropolis of Dermech (also in Carthage) the orientations of the tombs show an azimuth preference between 120° and 160°. Finally, the Maltese shaft tombs and burial chambers studied by Ventura (2000) do not show a definite solar or lunar pattern, but a clear preference for the approximate north-south direction.

The importance of the orientation in Punic ritual is documented in a stone inscription found in the zone of Salammbo in Carthage. This stone was an offering placed in a sanctuary dedicated to Baal-Hammon. The text indicates explicitly that the stone was oriented with its front side to sunset and its back side to sunrise (Xella 1991: 48). An additional interesting archaeological indication was obtained by L. Carton, who found that in an open sacred area in Thuburnica (Sidi-Ali-Bel-Kassem, Tunisia) all Neo-Punic stelai were oriented to the east (Leglay 1961: 276). This east-west arrangement of stelai and stone offerings is consistent with the general orientation pattern of Punic religious and funerary remains.

Esteban et al. (2001) have measured the orientations of a large sample of Roman and pre-Roman temples in Morocco, Tunisia, and Libya. These authors have found that the Roman temples, built mostly in the second and third centuries AD, show a definite random distribution of orientations and a lack of correlation between the dedication of the temples and their orientation, except in the case of those dedicated to Saturn (Fig. 3). From the impressive amount of iconographic and archaeological data analysed by Marcel Leglay (1961; 1966), continuity is demonstrated between the cult of the Punic Baal-Hammon and Saturn in Roman times from ancient Numidia to Mauretania, also the strong solar character of this god. On the other hand, the inscriptions of the sanctuaries at Thinnisut and El Hofra are further indications of the equivalence of the Punic pair Baal-Hammon/Tanit and the Roman Saturn/Caelestis. Leglay (1961; 1966) found that most of the temples dedicated to Saturn show the following interesting features: They are oriented with their entrances toward the east; they are located on high places outside the cities; and they are built over or very near previous pre-Roman sacred sites. In Table 2 and Fig. 3 we show the orientation of Roman North African temples dedicated to Saturn, including those directly measured by Esteban et al. (2001) and those measured from plans published by Leglay (1966), Berthier and Charlier (1955), Morestin (1980), and Hurst (1999). From Fig. 3 we can see that most of the religious buildings dedicated to Saturn are oriented toward the range of azimuths where the sun (or the moon) rises during the year. All these results suggest that the solar and/or lunar (perhaps, generally speaking, astral) aspects of the supreme North African god were maintained from Punic to Roman times.

Temple	Azimuth (°)	Declination (°)	Notes
Mactar	68 ± 1	+18 ± 1	measurements by Esteban et al. (2001)
Simithus	94 ± 1	−2 ± 1	measurements by Esteban et al. (2001)
T. Majus	53 ± 1	+30 ± 1	temple No. 29 (Esteban et al. 2001)
T. Majus	134 ± 1	−32 ± 1	temple No. 30 (Esteban et al. 2001)
Thugga	67 ± 1	+18 ± 1	temple No. 38 (Esteban et al. 2001)
Thugga	111 ± 1	−16 ± 1	temple No. 32 (Esteban et al. 2001)
Volubilis	94 ± 1	+1 ± 1	measurements by Esteban et al. (2001)
Timgad	~ 89	~ +1	from map (Leglay 1966)
Tiddis	~ 87	~ +2	from map (Leglay 1966)
El Hofra	~120	~ −24	from map (Berthier and Charlier 1955)
Hippona	~ 91	~ −1	from map (Morestin 1980)
Carthage	~ 99	~ −7	temple at the tophet, from map (Hurst 1999)

Table 2. Orientations of North African temples dedicated to Saturn.

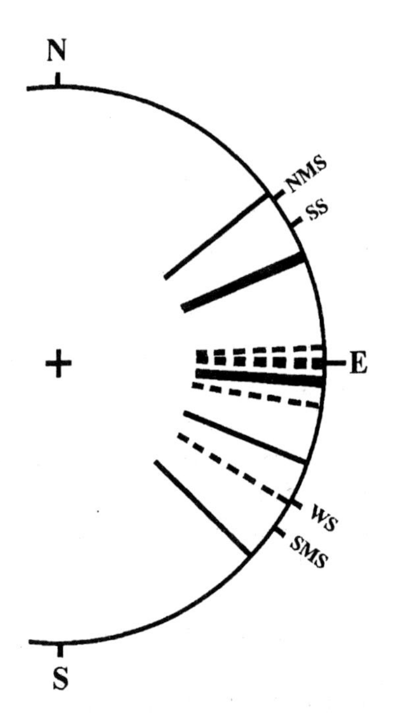

Fig. 3. Orientations of Roman temples dedicated to Saturn from all North Africa. Continuous lines: direct measurements; dashed lines: orientations obtained from published plans. See text for references. SS and WS mean summer and winter solstice; NMS and SMS mean northern and southern major standstills of the moon.

Eclipses over Carthage

Recently Léo Dubal (2000) has made a very suggestive conjecture connecting the almost total eclipse that occurred over Carthage on April 30, 463 BC, with three great changes in Punic culture: the starting point of the period during which the crescent moon was systematically engraved in the protective (upper) position about the solar disc (from the middle of the sixth century BC to the Roman conquest in 146 BC), the promotion of the goddess Tanit before Baal-Hammon in the religious inscriptions of Carthage (Xella 1991:42), and the beginning of the territorial expansion of Carthage.

Although the direct influence of this remarkable celestial event on those historical changes cannot possibly be proven, this hypothesis aroused our curiosity and led us to perform a somewhat more detailed study of the characteristics of that eclipse.

The map of the visibility of the eclipse in the western Mediterranean is shown in Fig. 4, where we have included the position of important settlements of Punic influence as well as the band of totality of the eclipse of April 30, 463 BC. The percentage of obscuration of the solar disc in each of the cities of Punic influence at the maximum extent of the eclipse and other relevant data are shown in Table 3 (100% of obscuration means that the eclipse is total). All the results have been obtained making use of the commercial computer program SKYMAP PRO7 (see http://www.skymap.com).

City	% of obscuration at maximum	Time (UT) of the maximum	Duration of totality
Carthage	99.3	12:31	-
Utica	100	12:31	1m 58s
Motya	100	12:35	3m 23s
Tharros	96.2	12:28	-
Cirta	100	12:23	2m 49s
Tipasa	100	12:17	4m 40s
Ebussus	93.9	12:14	-
Gades	92.5	11:56	-
Lixus	96.2	11:55	-

Table 3. Relevant data of the eclipse on April 30, 463 BC.

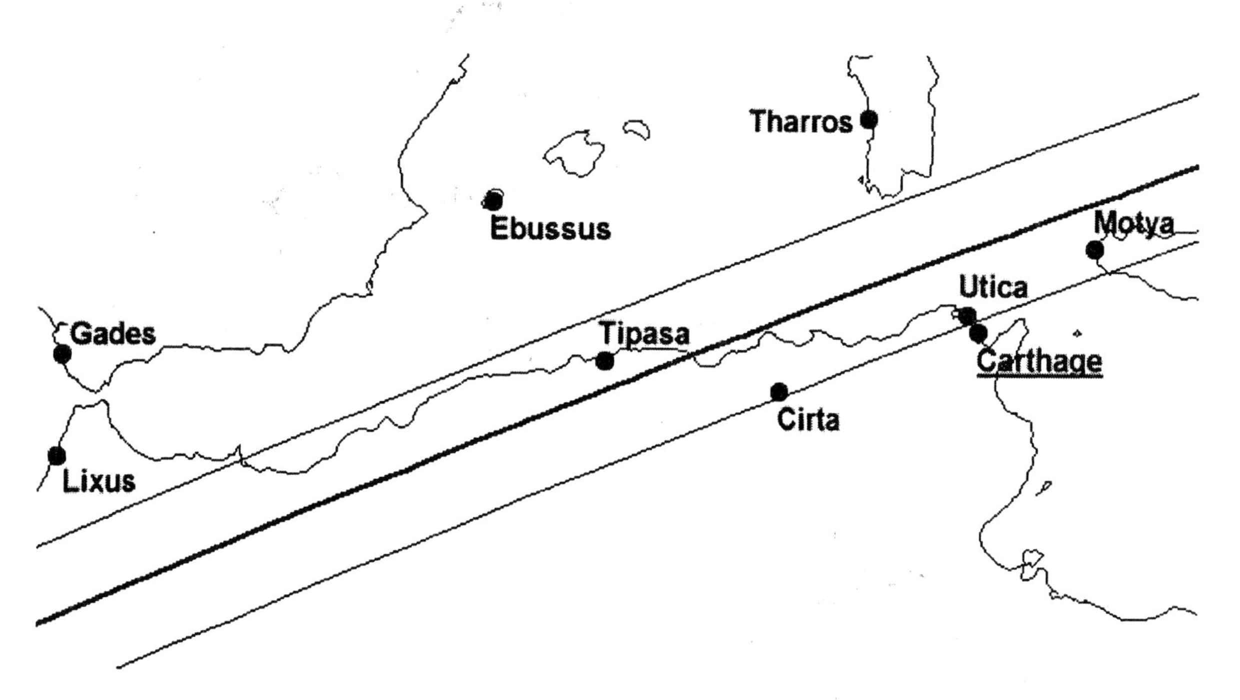

Fig. 4. Map of the visibility of the solar eclipse of 30 April 463 BC, in the western Mediterranean. The position of important settlements of Punic influence and the band of totality are included.

As can be seen in Table 3 and Fig. 4, the eclipse was observed under very good conditions (total or almost total) in most of the Mediterranean territory of Punic influence. Other remarkable characteristics of the eclipse are the following: a) It occurred 1h 17m after midday, at 60° of altitude, and in the conspicuous constellation of Taurus, just between the Hyades and Pleiades (Fig. 5); b) Mars and Venus were located almost equidistant (about 13°) on opposite sides of the sun; c) A very small crescent, covering only 0.7% of the sun disc, was observed below the moon disc (Fig. 6) at the maximum of the eclipse in Carthage. The striking features of the eclipse make it very likely that it would have produced a profound impact on the Punic observers. Moreover, the news that this eclipse was observed in almost all the Punic world and that it became total in cities so close as Utica (only 30 km to the north!) probably led the Punic priests and rulers to ask themselves, the gods, and the oracles about the meaning of this unique celestial omen.

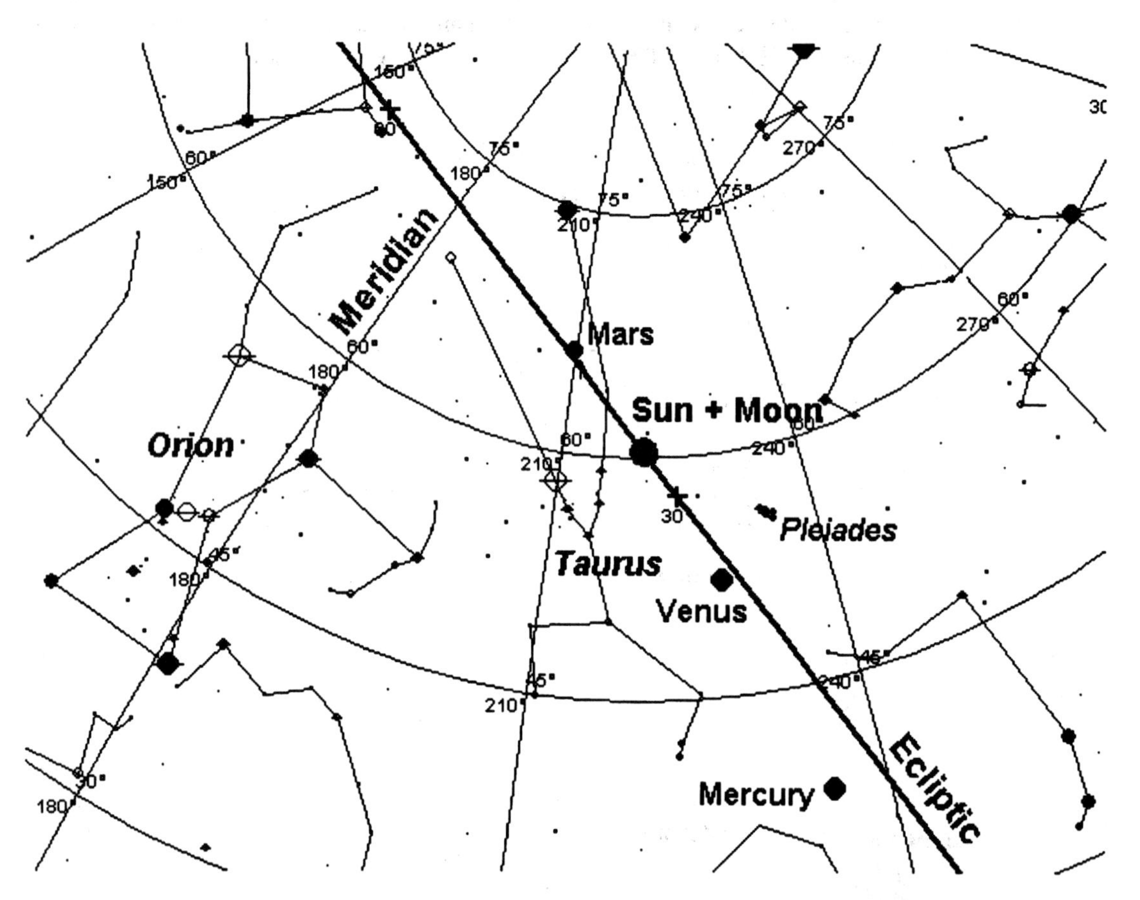

Fig. 5. Map of the sky as seen from Carthage at the maximum of the solar eclipse of April 30th, 463 BC. The grid is in horizontal coordinates, the zenith is at the top.

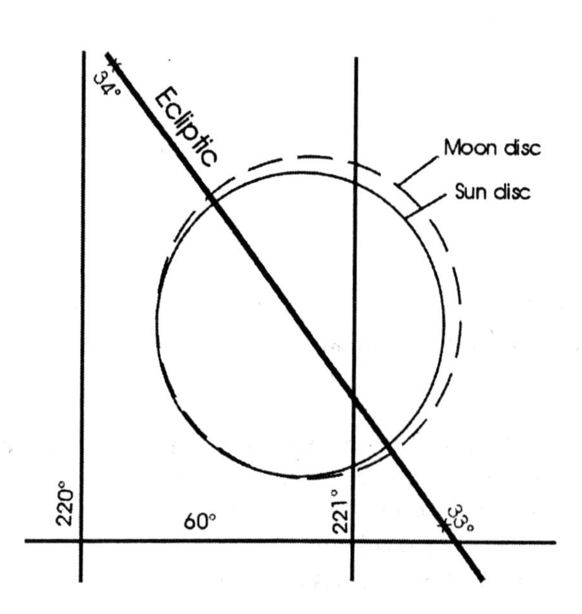

Fig. 6. Detail of the relative positions of the sun and moon discs at the maximum of the solar eclipse of April 30th, 463 BC, as seen from Carthage. An extremely narrow crescent of the sun is seen below the moon disc. The grid is in horizontal coordinates.

We have explored the occurrence of other relevant solar eclipses during the period of the existence of the Punic state, from the beginning of the eighth century BC to the Roman conquest in 146 BC. We have found that only two total eclipses and no interesting partial ones (with a percentage of obscuration larger than 95%) took place over Carthage in that period. The total eclipse on February 10, 765 BC, occurred during sunrise and it was not well observed in the western Mediterranean. The second eclipse, which occurred on August 15, 310 BC, is more interesting because its band of totality had a similar geographical distribution as the eclipse of 463 BC (it was well observed in Punic territory except for Tripolitania) and it occurred at an altitude of 28° over the Carthaginian skies. It is curious and suggestive that this second eclipse coincided precisely with the war expedition of Agathocles, tyrant of Syracuse, who inflicted a great defeat on the Carthaginian army in African lands. There were important changes in Punic religion coincident with this military event. Firstly, there was a general return to traditional ritual and an official 'reconciliation' with the old cult of Baal-Hammon, with the sacrifice of several hundreds of young children of the most important families. Secondly, there was a re-initiation of the sending of gold and other sacred offerings to the temple of Melqart in the mother-city of Tyre, an ancient custom that had been almost neglected for a long time (Huss 1993: 124-126). Did the total eclipse of 310 BC contribute to promote these religious changes? We will never know, but the coincidence of the two most relevant solar eclipses visible from Carthage in the whole Punic era with two periods of important religious changes suggests that this could be a reasonable hypothesis, but unfortunately this is impossible to prove. One other question remains to be answered: Were those days of the eclipses cloudy over Carthage?

*Instituto de Astrofísica de Canarias, 38200 La Laguna, Tenerife, Spain, cel@ll.iac.es

References

Alaoni, M. K. 2000. A propos de la chronologie de la nécropole rurale d'Aïn Dalia Lekbira (region de Tanger, Maroc). In *Actas del IV Congreso Internacional de Estudios Fenicios y Púnicos*, vol. 3, ed. M. E. Aubet and M. Barthélemy, 1185-1195. Cádiz.

Baradez, J. 1969. Nécropole orientale côtière de Tipasa de Maurétanie. *Antiquités Africaines* **3**, 83-114.

Belmonte, J. A. 1999. *Las leyes del cielo*. Madrid.

Belmonte, J. A., Estaban, C., and Jiménez González, J. J. 1998. Mediterranean archaeoastronomy and archaeotopography: Pre-Roman tombs of Africa Proconsularis. *Archaeoastronomy* (Supplement to the *Journal for the History of Astronomy*) **23**, S7-S24.

Berthier, A. and Charlier, R. 1955. *Le sanctuaire punique d'El Hofra à Constantine*. Paris.

Dubal, L. 2000. The riddle of the protective crescent in Punic votive art. In *Actas del IV Congreso Internacional de Estudios Fenicios y Púnicos*, vol. 2, ed. M. E. Aubet and M. Barthélemy, 583-591. Cádiz..

Esteban, C., Belmonte, J. A., Perera Betancourt, M. A. Marrero, R., and Jiménez González, J. J. 2001. Orientations of Pre-Islamic temples of Northwest Africa. *Archaeoastronomy* (Supplement to the *Journal for the History of Astronomy*) **26**, S65-S84.

Hurst, H. 1999. *The Sanctuary of Tanit at Carthage in the Roman Period* (*Journal of Roman Archaeology*, Supplement 30). Portsmouth.

Huss, W. 1993. *Los cartagineses*. Madrid.

Lancel, S. 1994. *Cartago*. Barcelona.

Leglay, M. 1961. *Saturne Africain. Monuments*, vol. 1. Paris.

Leglay, M. 1966. *Saturne Africain. Monuments*, vol. 2. Paris.

Merlin, A. 1910. *Le sanctuaire de Baal et de Tanit près de Siagu*. Paris.

Morestin, H. 1980. *Le temple B de Volubilis*. Paris.

Niemeyer, H. G. 2000. Un nuevo santuario de la diosa Tanit en Cartago. In *Actas del IV Congreso Internacional de Estudios Fenicios y Púnicos*, vol. 2, ed. M. E. Aubet and M. Barthélemy, 635-642. Cádiz..

Picard, G. Ch. 1954. *Les religions de l'Afrique antique*. Paris.

Ribichini, S. and Xella, P. 1994. *La religione fenicia e punica in Italia*. Roma.

Stieglitz, R. R. 2000. The Phoenician-Punic Calendar. In *Actas del IV Congreso Internacional de Estudios Fenicios y Púnicos*, vol. 2, ed. M. E. Aubet and M. Barthélemy, 691-695. Cádiz.
Tusa, V. 2000. Il santuario fenicio-punico di Mozia, detto di "Cappiddazzu". In *Actas del IV Congreso Internacional de Estudios Fenicios y Púnicos*, vol. 3, ed. M. E. Aubet and M. Barthélemy, 1397-1417. Cádiz.
Ventura, F. 2000. Orientations of the Phoenician and Punic shaft tombs of Malta. In *Oxford VI and SEAC 99. Astronomy and cultural diversity*, ed. C. Esteban and J. A. Belmonte, 59-63. Santa Cruz de Tenerife.
Xella, P. 1991. *Baal-Hammon*. Rome.

Orientation of some dry stone monuments: 'V shape' monuments and 'goulets' of the Immidir Mountains (Algeria)

Yves Gauthier* and Christine Gauthier*

Abstract
This work is part of a general study of lithic constructions of central Sahara, with a focus on their orientations. We report here on two different types of monuments of the Immidir massif (Algeria), namely 'V shape' monuments and 'goulets'. The former is specific to the area, whereas the latter is present from central Sahara to western Sahara. We find that the azimuthal distribution of their main directions lies almost exactly within the limits of the position of the sun at the winter and summer solstices, supporting the hypothesis of a sunrise-based alignment, as proposed for other monuments. This tradition, probably dating back to the 6th millennium BC, was still in use around 1500 BC, or not much later.

Introduction
Stone monuments are far from rare in the Sahara regions, and architectural types are quite numerous. These monuments, which in many, but not all cases had a funerary function, span millennia and are among the most spectacular remains from neolithic and prehistoric populations. Along with paintings and engravings they are important cultural markers, typical of a group and a time, some having a wide geographical distribution, while others are restricted to comparatively small areas. Visitors travelling through the central Sahara may quickly realise that, in spite of some striking differences, a large proportion of these monuments have a tendency to be oriented roughly eastwards. Such a trend is quite well documented in detailed studies of the so-called 'keyhole' monuments of the Fadnoun Plateau, Algeria (Savary 1966) or of NE Ténéré, Niger (Paris 1996), of the 'platform cairns with 2 arms forming a V' and finally of 'platform cairns with an arm running eastwards' (Gauthier and Gauthier 2000). Whatever the actual object (celestial or not) or criterion used for the alignment of these monuments, it is apparent that the distribution of the orientations is rather narrow. Statistics based on more than 150, 47, and 26 specimens respectively shows that the main directions of these monuments lie most often between the azimuths of the rising sun at the summer and winter solstices and that, when they do not, the direction is shifted slightly towards the south.

For those constructions with meaningful orientations, is this trend general or do different rules apply to monuments of differing regions? Work is in progress on a dozen monument types in South Libya and South Algeria with an emphasis on the Immidir region (Algeria). In the present study we focus on two particular types of monuments that are frequent in the latter region: 'V shape' monuments ('Vs' in the following) and 'goulets'. Besides their obvious architectural differences they differ also in respect of their distribution. They do, however, show rather common rules regarding orientation.

Distribution and Location
All the monuments considered below were observed exclusively in the mountains lying roughly ENE of Arak (Fig. 1a-b), between 25°20' and 25°50' N and 3°50' and 4°35'E. This does not constitute an exhaustive survey of the monuments of the massif. Indeed cars cannot penetrate the region, and in spite of the many kilometres made on foot (more than 1500), many places are still yet to be visited and many constructions remain to be discovered. Nevertheless, the number of each type of monument is statistically significant and permits the drawing of some preliminary conclusions as to their plan, location, distribution, and orientation. So far we have observed 44 'Vs' and 32 'goulets'. A full description of the monuments will be given elsewhere with a discussion of the typology and variations.

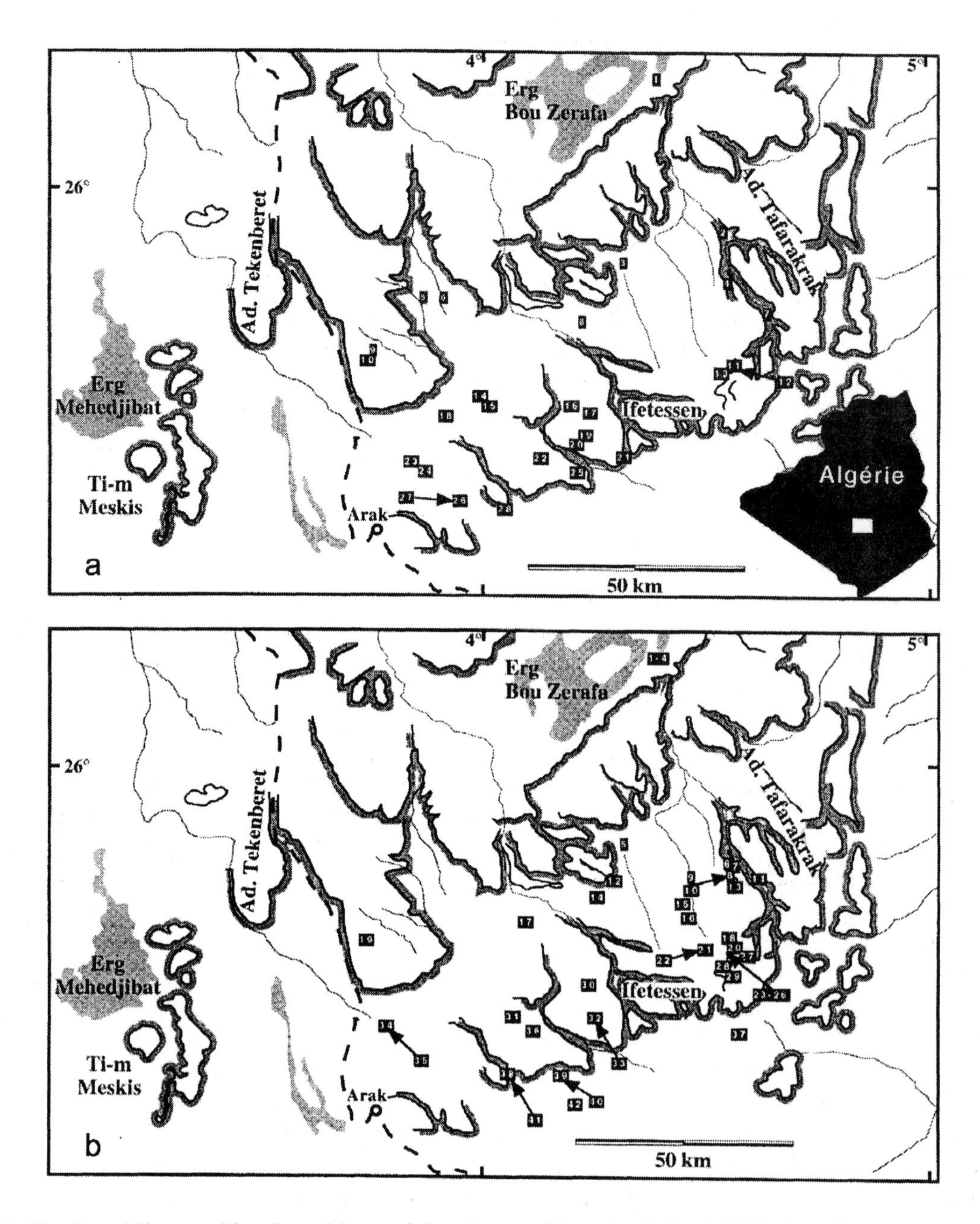

Fig. 1. The Immidir massif and positions of the observed 'goulets' (a) and 'V shape' monuments (b).

The maps of Fig. 1 illustrate a wide distribution of both types of monument spread over the massif. However the respective densities of the two types are not really similar. We note for instance that approximately half of the 'Vs' are located in the Ifetesen sector or close to it. This uneven density, or clustering, may result from the morphology of the mountains and the availability of water resources that doubtlessly influenced the settlements of the builders. It may also be due to the incompleteness of the survey due to logistics, as mentioned above.

These monuments were generally built on flat areas. However owing to the morphology of the region—mainly mountains—it is not rare to find them on slopes, in hollows, or at the bottom of hills. In other words, while some actually lie in open areas—plains, tops of high plateaus—with no apparent horizon, many occur in hilly areas. For the latter the consequence is that any celestial object (e.g. sun, moon) taken as a reference for the orientations of the monuments will appear later and/or will disappear sooner than for the former. This implies that the local azimuth of the rising sun with respect to north, for instance, will be shifted a few degrees toward the south.

In several places 'Vs' and 'goulets' are separated by only a few meters, showing that they are not mutually exclusive. This suggests various possibilities: 1) two cultural groups actually coexisting, 2) successive groups, 3) one group only, with several rites.

At this time we lack ^{14}C dating and, unfortunately, have no example so far of superposition that might indicate that one type is older than the other.

The presence of 'Vs' is not surprising at all. Similar monuments were observed in the neighbouring regions to the east, to the west, and to the south (Voinot 1904; Monod 1932; Savary 1955; Reygasse 1950) and, more generally, they are widely spread in Central Sahara. The situation is quite different for the 'goulets'; none have yet been discovered elsewhere in Central Sahara and they seem to be quite specific to this massif (Gauthier et al. 1998)! This fact may be related to the presence of particular layers of paintings that apparently have no equivalent in the other nearby regions with well-documented rock art traditions (Tassili-n-Ajjer, Tefedest, Ahaggar, Ahnet). However nothing so far allows the attribution of the latter monuments and these particular paintings to the same group. Still, concerning the 'goulets', we wish to mention that similar constructions have been discovered in Western Sahara (Monod 1948), some 1500 km west of Immidir without any other (known) example in between.

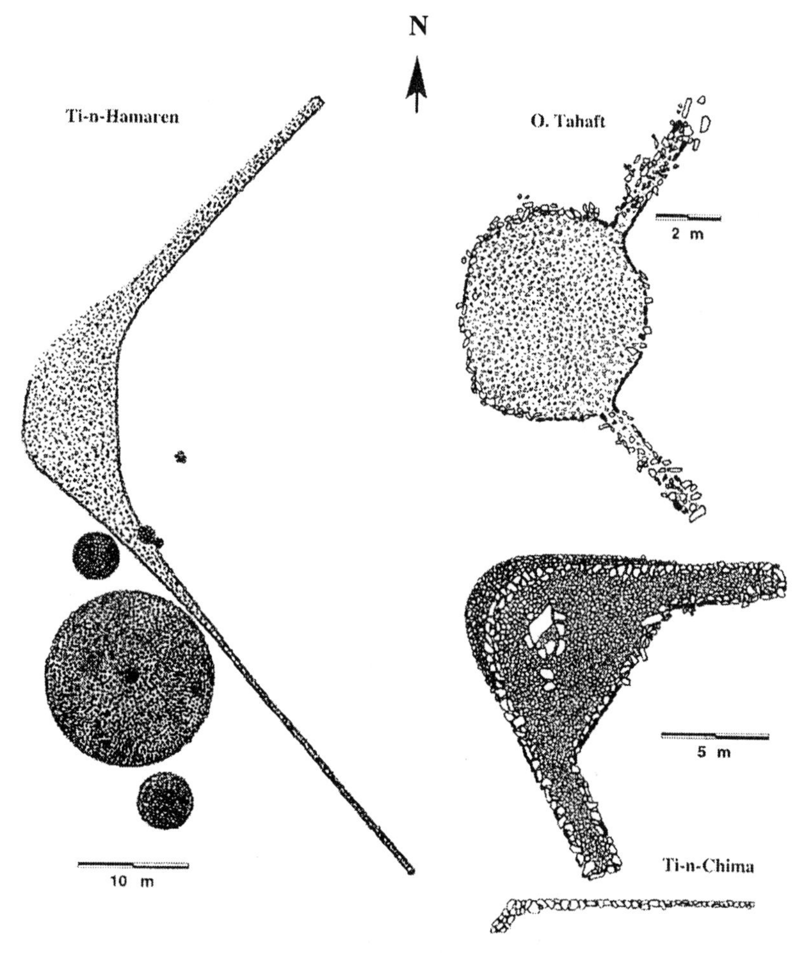

Fig. 2. Examples of 'V shape' monuments of Immidir. Sometimes the border is marked by upright slabs or bigger blocks. The construction of Ti-n-Hamaren is associated with other monuments (disks) of unknown chronology. The burial in the monument of Ti-n-Chima was looted, as indicated by the large slabs lying in the middle.

Description

'V shape' monuments The 'V shape' monuments we are discussing here consist of a main part with two arms or antennas running approximately NE and SE respectively (Fig. 2). These antennas, which form a 'V', may extend over quite a long distance (~ 70 m) or be extremely short (~ 2.5 m), if not almost absent. The shape of the main part is variable, from roundish to crescent-like. Most often the western side is a portion of a circle prolonged by the antennas, while on the eastern side, between the antennas, the border line is gently curved or straight. In the case of the most elaborate constructions, slabs fixed upright in the ground mark the border. The inner part of the monument is made of stones or blocks of variable dimensions, generally without any particular arrangement. The main part is usually somewhat flat (h $\leq$ 0.5 m above ground), sometimes with a central circular area 10 to 20 cm higher than the rest, marking the position of the burial (Fig. 3). We have also included in our study some monuments with very short antennas and a similar outline but with a true tumulus in the centre (h ~ 1-2 m).

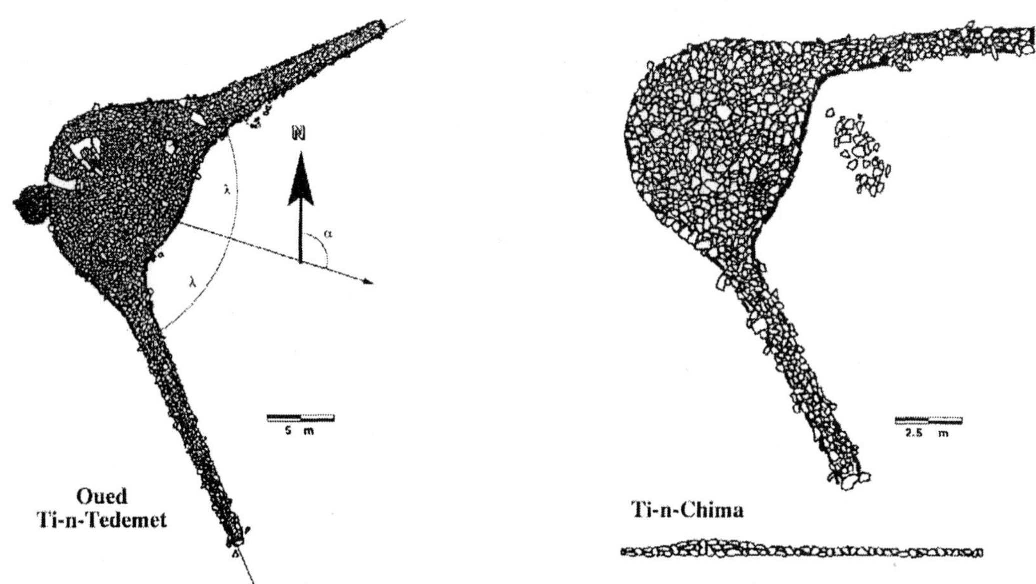

Fig. 3. Definition of the orientation (α) of 'V shape' monuments. The cut of the Ti-n-Chima monument, along the bisector of the angle of the antennas, shows the position of the burial (higher part on the left).

'Goulets' The 'goulets' are enclosures (up to 100 m long), generally elliptical or roundish, divided into two fairly symmetrical portions by a central corridor (Fig. 4). This corridor, roughly oriented east-west, is terminated on its western end by a tumulus that is tangential to the enclosure or overlaps with the walls. In many instances the enclosure and the corridor are outlined by walls 0.5 to 1 m high and 1 to 3 m wide. These walls are built with irregular and non-ordered blocks and sometimes are made of two lines of flat slabs inclined and facing each other. The corridor (2 to 4 m wide) may be outlined by two external walls, by a paved area, or by a combination of both. Some rare 'goulets', the smallest ones, are (almost) entirely paved and their corridors may be so narrow as to be nearly non-existent.

Orientation

For the 'Vs' the orientation (α) is defined by the direction with respect to north of the bisector of the angle formed by the arms (Fig. 3). Quite naturally the orientation of a 'goulet' (β) is given by the direction of the corridor (Fig. 5).

The direction of the arms for the 'Vs' and that of the corridor for the 'goulets' were measured with a magnetic compass, whereas the angle of the apparent horizon was measured with a clinometer, both with a precision of ± 1°. From one year to another, variations in the directions of 1°, exceptionally 2°, were noted for some monuments. This main source of error is due to the difficulty in defining precisely the reference points at both ends of the corridor (one in the centre of the tumulus) or of the arms. In the following the magnetic declination (average ~ -1.8°) was accounted for using the data specified on the IGN maps (NG-31-XI, *Ifetessène* and neighbours).

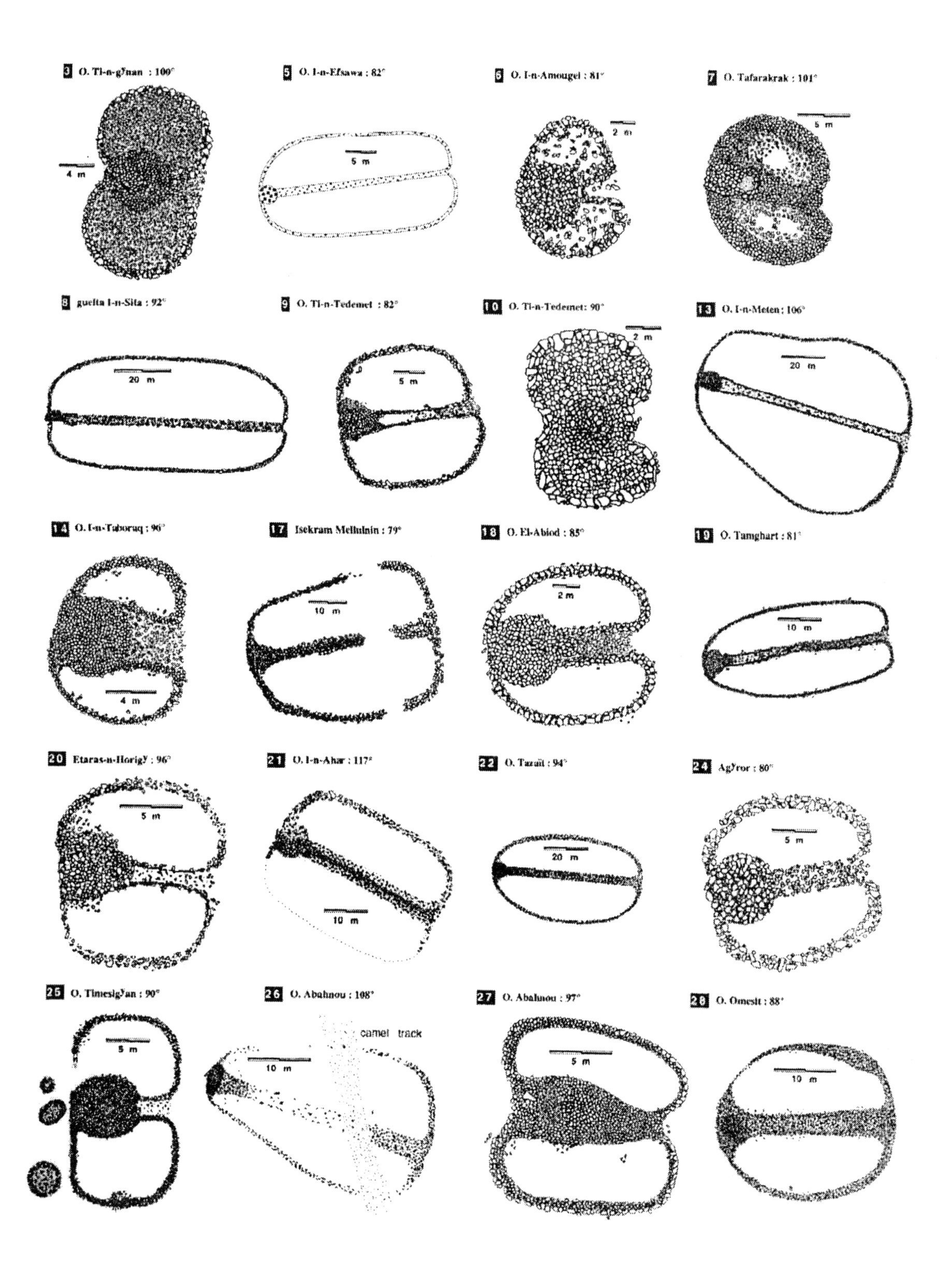

Fig. 4. Location and azimuth of the corridor of the 'goulets'. The numbers (white) are those of Fig. 1. Orientations, dimensions, and proportions are highly variable; on the smallest examples the corridor is shorter than the breadth, while the opposite tends to be the case for the biggest constructions. Note the complete paving of some monuments and of many corridors. North is to the top of the page.

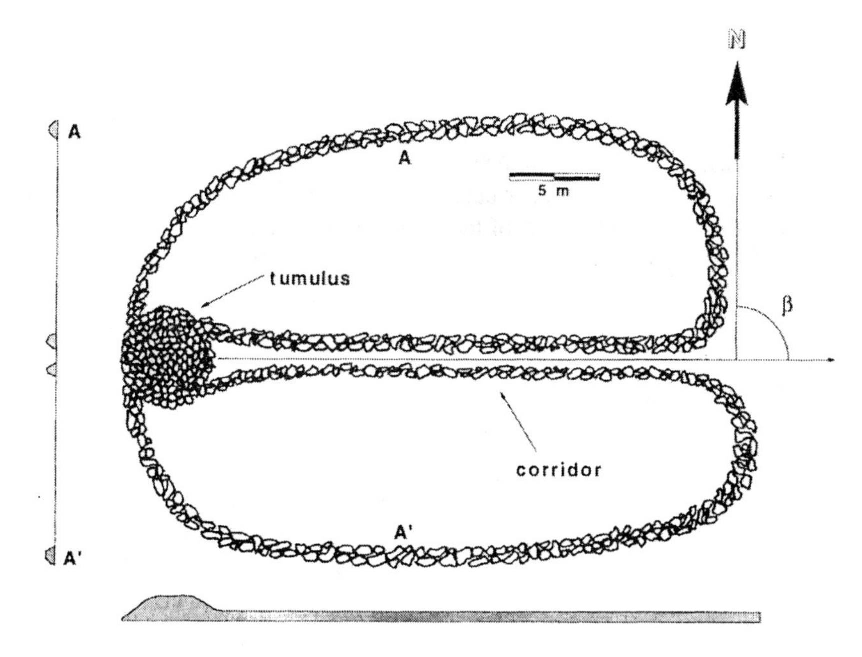

Fig. 5. Definition of the orientation (β) of the 'goulets'. The example is from Oued I-n-Taboraq. The two additional profiles correspond to cuts through lines AA' and to a projection in the vertical direction parallel to the corridor, showing the height of the different components.

We measured the orientations of 31 'Vs', out of 44. If we plot the azimuth α (Fig. 6), we find that these monuments consistently open to the east with an extremely narrow distribution ($66° \leq \alpha \leq 118°$); all α values, except for two, lie strictly between the limits of the rising sun (at sea level) at the summer and winter solstices, 63° and 116° respectively. For the positions of the sun at the solstices during the Neolithic period we have considered the values proposed for the Fadnoun (Algeria) at the same latitude (Savary 1966). The two monuments that have a direction outside these limits show a very weak southward shift of the azimuth ($\alpha = 116°$ and 118° respectively). This 2° shift might not be significant owing to the precision of the measurements and the slightly different possible positions for the reference points.

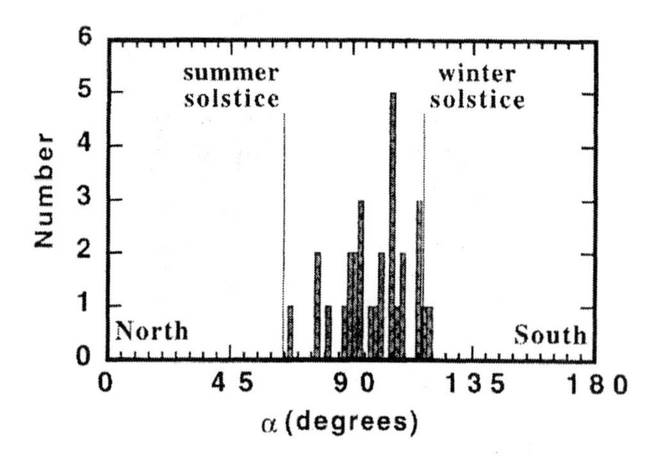

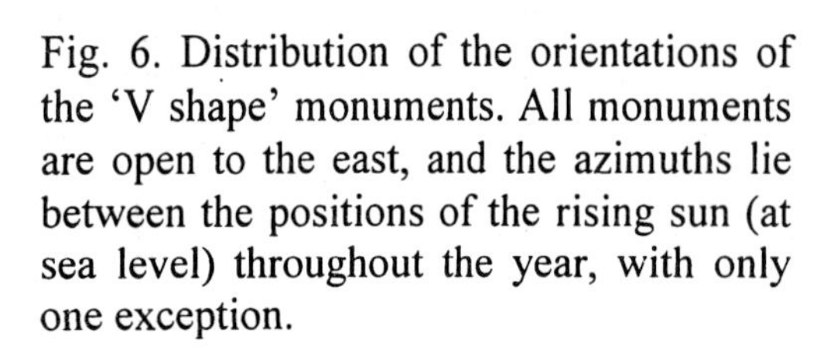

Fig. 6. Distribution of the orientations of the 'V shape' monuments. All monuments are open to the east, and the azimuths lie between the positions of the rising sun (at sea level) throughout the year, with only one exception.

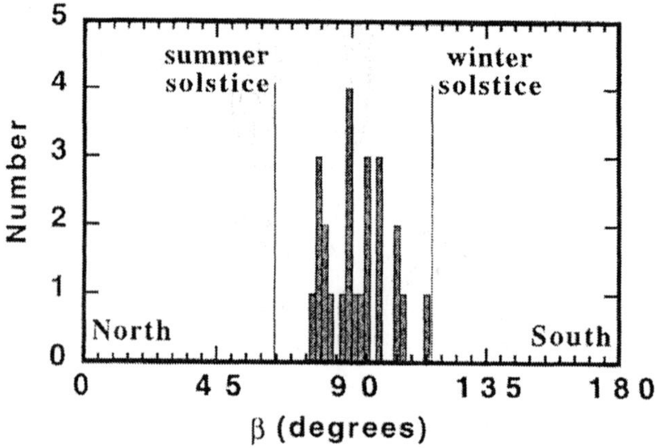

Fig. 7. Orientations (β) of the 'goulets'. The directions of the corridors are consistently between the azimuths of the sun at the winter and summer solstices.

Quite similar conclusions may be drawn from Fig. 7 concerning the 28 'goulets' (out of 32) for which the orientations were measured. Here too all azimuths lie within the limits of sunrise over the year. Their distribution is even narrower ($73° \leq \beta \leq 115°$) than for the 'Vs'.

Discussion

In short, the orientation of these constructions is not random and must be due to rather strict rules. This is not a particular case as quite similar distributions have already revealed for other monument types. At this point it is important to situate the above results in the larger context of the Sahara and Time. Rising-sun alignment was proposed first for 'keyhole' monuments of Fadnoun, Algeria (Savary 1966) and Niger (Paris 1996).

For the Fadnoun, which is as mountainous a region as Immidir, Savary observed that while a large proportion of the 'keyhole' monuments had their orientations strictly within the position of the solstices about 10% were shifted southwards, never northwards. The author concluded that the orientation of the constructions was based on the rising sun, the southwards shift being fairly well explained by the presence of relief to the east, as explained above. The conclusion—southward shift—would be the same if the reference were the sunset. Concerning the 'keyhole' monuments of Niger (Paris 1996), built in a flat area, the distribution happens to be more precise and well within the solstitial limits.

We note that out of 59 measured 'Vs' and 'goulets', only one monument has an azimuth shifted south of the solstitial limit in spite of their location, exclusively in mountains. The apparent horizon in the east was measured for some monuments only: the angular height of the highest point is mainly ≤ 3° but may reach 8°. However it is certainly much larger for other monuments for which we have not collected this parameter. This should result in a southwards shift of the azimuth whenever the rising sun is the reference. Generally speaking, those monuments that have larger azimuths—i.e. close to the winter solstice limit—are those with the highest apparent horizon to the east. In other words, the distribution might have been even narrower if the monuments had been established on totally flat plains and not in the mountains. *This clearly needs to be confirmed.*

Nevertheless, owing to this narrow distribution about east, the rising sun appears a highly probable reference for the builders. We have a slight preference for the rising sun rather than for sunset because of the opening of the 'Vs' and of the corridors of the 'goulets' to the east. An additional argument in favour of the rising sun is that the monuments are much more often on eastern slopes—thus facing the rising sun—than on western slopes. Finally, in the limited number of cases for which we took note of the angular height of the horizon, we find much larger angles in the direction of sunset rather than of sunrise. As a consequence an orientation with respect to sunset—instead of sunrise—probably would not fit, as the declination for the celestial object taken as reference would fall well outside the limitations of the sun.

In fact such narrow azimuthal distributions (almost) within solar limits are more common: 'V shape' monuments, 'platform cairns with an arm running eastwards' (Gauthier and Gauthier 2000) and 'L shape' monuments *stricto sensu* (Gauthier and Gauthier 2002) of Fezzan (Libya) are also directed eastwards. And the list is not complete!

To strengthen the point we have gathered the results for the 'platform cairns with an arm running eastwards', for the 'goulets', and for the 'Vs' of both Immidir and Fezzan that are at the same latitude, 24°30' ± 1°, i.e. a total of 132 constructions (Fig. 8). The distribution curve is centred close to the East (95°), with only eight monuments falling outside the solstitial limits. The two monuments shifted northwards are not more than 3° beyond the limit, which is acceptable considering the precision. It is striking to notice that this distribution is very similar to that of the 'keyhole' monuments of Fadnoun situated at the same latitude. Both curves are slightly shifted to higher azimuths with respect to the east, and those monuments that are not within the solstitial limits are similarly in the SE quadrant, in very good agreement with the above hypothesis of a sunrise alignment in the presence of a non-flat horizon.

This suggests that a more detailed analysis is needed to account properly for the apparent horizon; such an analysis might reveal even stricter rules. In any event it is already clear that, sun based alignment or not, the same rules have prevailed over very large distances since, like those from Central Sahara, the 'Vs' of western Sahara have also an eastward orientation. The presence of these rules in both locations is indicative of the diffusion, if not of people, at least of some funerary customs or rituals.

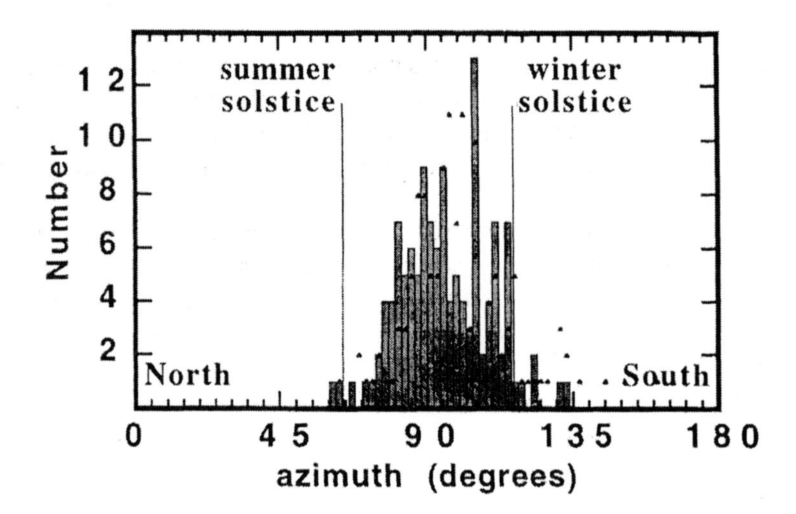

Fig. 8. Orientations of 'Vs' and 'goulets' of Immidir, Algeria, and of 'Vs' and 'platform cairns with an arm running eastwards' of Fezzan (Gauthier and Gauthier 2000; 2002). The dots represent the 'keyhole' monuments of Fadnoun (Savary 1966).

However, it would be wrong to generalise to all Saharan lithic constructions for the following reasons: 1) 'crescents' or 'triple crescents' of Messak, Libya, do not fulfil the above criteria (Gauthier and Gauthier 2000); 2) work in progress on *Tentes de Fatima/anses de panier* yields a much wider azimuthal distribution; 3) some 'crescents' of South Algeria do show orientations to both the east and the west; 4) finally, it is worth noting that the 'goulets' of western Sahara may have various orientations; one is facing west, β =280° (Monod 1948). This non-exhaustive list indicates that rites and beliefs were numerous, resulting from different populations and influences through the ages, in agreement with what is known from burials and from rock art.

As for the date of the origin of sun-based orientations it is worth recalling that 'keyhole' monuments were erected between 4700 and 3000 BC (Paris 1996), while the 'Vs' are much younger, ~ 1500 BC (Collectif 2003). It may be postulated that this tradition, which survived for millennia and probably until the onset of Islam, is even older and *began sometime during the 6th millennium BC*, which would have provided sufficient time for *the observed diffusion all over central and western Sahara.*

*264, rue de la Balme, F-38950 Saint Martin le Vinoux, France, yves.gauthier@grenoble.cnrs.fr.

Acknowledgements
We wish to thank the director of the Office du Parc National de l'Ahaggar, who allowed us to visit the region and to realise this work. We acknowledge the help of Jacqueline and Jean Claude Friquet, Henri Lahuppe, Jean Claude Piermet, and Jean Rodière, during the fieldwork. Warm thanks are also due to Faragi Ag Fassi, Cheikh Ag Fassi, El Faqi Ag Fassi, Attlagh Bacadi, and Rali Khamdani; without their thorough knowledge of the region, we would not have been able to penetrate the massif and to discover these monuments. Line drawings are by the authors.

References

Collectif 2003. Sand, stone, and bones. The archaeology of death in the Wadi Tanezzuft Valley. In *Arid zone archaeology* (Università La Sapienza di Roma monograph 3 and Department of Antiquities of Libya), ed. S. di Lernia and G. Manzi. In press.

Gauthier, Y., Nöther, W., and Lluch, P. 1998. Monuments de L'Immidir (Algérie). *Sahara* **9**, 143-148.

Gauthier, Y. and Gauthier, C. 2000. Orientation et distribution de divers types de monuments lithiques du Messak et des régions voisines (Fezzân, Libye). *Sahara* **11**, 87-108.

Gauthier, Y. and Gauthier, C. 2002. Monuments à antenne en "L" ou apparentés. Une originalité du Fezzan? Architecture, orientation et distribution. *Sahara* **13**, 136-147.

Monod, Th. 1932. *L'Adrar Ahnet. Contribution à l'étude archéologique d'un district saharien.* (Travaux et Mémoires de l'Institut d'Ethnologie 19). Paris.

Monod Th. 1948. Sur quelques monuments lithiques du Sahara occidental. *Actas y Memorias de la Sociedad. Española de Antropología., Etnografía. y Prehistoria* **23**, 12-35.

Paris, F. 1996. *Les sépultures du Sahara nigérien, du néolithique à l'islamisation.* 2 vol. (Etudes et thèses de l' Institut Français de recherche scientifique pour le développement en cooperation). Paris.

Reygasse, M. 1950. *Monuments funéraires préislamiques de l'Afrique du Nord* (Arts et Métiers Graphiques). Paris.

Savary, J. P. 1966. *Monuments en pierres sèches du Fadnoun (Tassili n'Ajjer)* (Mémoires du Centre de recherches anthropologiques, préhistoriques et ethnographiques 6). Paris.

Voinot, Lt. 1904. A travers le Mouydir. Rapport de tournée. *Renseignements Coloniaux* (Supplement to the *Bulletin du Comité de l'Afrique française*, Annuaire et mémoires du Comité d'études historiques et scientifiques de l'Afrique occidentale française) **9**, 205-213 and **10**, 243-251.

Human beings and the stars: An anthropo-astronomical look at archaeology

Florin Stanescu*

Abstract
This paper presents a research project the intention of which is to study the spiritual universe of the Dacians and, more precisely, the way in which the great astronomical cycles have influenced its character. Also of interest is how the Dacians have represented these cycles, tracing this in both material and spiritual manifestations down to our times and in more than one geographical region.

General considerations
Trying to determine the ways of thinking of people who lived thousands of years ago, even in just one limited domain of their spirituality, is undoubtedly a difficult enterprise, and also a risky one. This is even more difficult when a number of relevant elements are partially or completely missing. The risk is more a matter of the interpretation of the surviving remains, especially when they are fragmentary. Even today different peoples in different places experience and interpret the same phenomena in very different ways. It is often with difficulty for us to understand or explain to ourselves the behaviour or experiences of people who today still live at the level of the Stone Age. With the help of the social sciences, e.g. anthropology, comparing and extrapolating from ethnographic analogies, we can try to understand even cultures that exist no longer. Here again there are risks, which have their origins precisely in our tendency to ascribe to people from other eras our own way of seeing things.

The research area
Five thousand years ago graves at Cernica near Bucharest were oriented towards sunrise. Today our churches have their altars orientated towards sunrise, and our icons are situated on the east walls of our houses. At sometime around the middle of this long period of time the Dacians themselves orientated their sanctuaries towards sunrise, the moon, or some stars. This means that the period of our investigations is necessarily very long, from the Neolithic until the present. It is our intention to achieve over this long period of time an overview of this type of manifestation in the cultures before, after, and during the time of the Dacians.

The study of the orientations of similar ancient sanctuary-like structures, or other constructions used for religious purposes, from the Neolithic until today throughout large areas of Europe, and also beyond, has shown a close relationship between the structures and the celestial bodies: the sun, the moon, and the stars of certain constellations. In practice this means searching in the field for alignments to these bodies at significant positions at the horizon, e.g. sunrise at the solstices and also for orientations towards bright stars, or even precise orientations towards the north-south direction from the great sanctuaries of the Dacians at Sarmizegetusa-Regia, Costesti, or Racos, specifically with respect to their central apses, altars, and series of columns and plinths. A series of assumptions must be kept in mind concerning the methods used by the builders to determine these alignments and also the astronomical information that they must have had for the purpose. It is very probable that relationships to the solstices were of great importance in the spiritual lives of these people. We also hope to show that a number of ancient writers have given us information about the special astronomical knowledge possessed by the Dacians. The existence of such orientations with respect to the monuments of the Dacians would confirm their participation in the general culture of their epoch and the existence of a close relationship between them and the great cycles of nature with respect to time and space.

In discussing the scientific dimensions of ancient knowledge, we will hold that there are three such dimensions: an empirical dimension deriving from observations in context, e.g. the alignments of monuments to stars; an analytical dimension built on the basis of logical analysis with the help of rudimentary mathematics, e.g. research experiments for establishing relations between a calendar and agrarian rituals; a thematic dimension, which concerns the elaboration of ideas. We will try to integrate the results of our investigation into one of these dimensions.

Archeoastronomy

The research domain The attempt to understand the ways in which different peoples in the past thought about themselves, about their environment, space, time, and so forth is made difficult by a number of factors, one of these being the absence of written documents, which is fundamental to communication in our present culture. This is a serious disadvantage to research in this domain, especially to that of archaeology and the history of astronomy. On the other hand, it is clear that the ancient astronomies of some cultures were not dependent exclusively on writing. Archaeoastronomy has developed as a new science in answer to the problems existing at the border between archaeology and astronomy. Many scholars consider it as belonging to the history of science, the history of astronomy in particular; others see it as belonging to archaeology. Archaeoastronomy is concerned especially with the determination of the orientation of ancient remains of specific kinds and also with the relationship of these remains to the astronomical phenomena of their location in order to be able to deduce the customs, behaviours, beliefs, and knowledge—particularly of astronomy—of the people who built them.

Characteristics and limits Archaeoastronomy, as defined by Aveni (1985), studies both written and unwritten evidence that refers to the astronomical practices of ancient civilisations. Unwritten evidence is constituted by material remains of civil and ceremonial architecture, and it is its design or shape, together with its ornamentation and relationships with the whole landscape, that should be researched.

In a traditional way archaeology is conceptualised as a study of different types of material remains such as walls, floors, or other elements of architectural structures, tools, pottery shards, and so on, which are considered as manifestations of ancient activities while, on the other hand, epigraphers and philologists analyse ancient texts. Because of their different origins these disciplines have evolved differently, establishing specific research goals and methodologies as well as particular criteria for the establishment of truth. Because of these disciplinary divisions some of the conclusions drawn from an archaeoastronomical approach may contribute to the research discussions of archaeologists, while others may eventually be part of a research agenda of historians.

The important question is to analyse how a simple alignment, defined as a line between two points in space, considered by archaeoastronomers as constituting a deliberate and significant orientation, can become meaningful for archaeologists. From this perspective astronomers tend to rely on statistical analysis of alignments to decide whether they were made intentionally or not. Where many different alignments are scattered around the horizon, it is important to judge the probability that they were made intentionally, and not fortuitously.

In their turn archaeologists will seek for a variety of possible culture-context influences in deciding which particular orientations of material features were meaningful. In other words they try to determine what were the symbolic (religious) or pragmatic (civil) considerations of making such and such orientations.

However the fascination with archaeoastronomy and its popularity has attracted laymen who are absolutely incompetent in this field. Thus on the one hand there are individuals who imagine that our ancestors were like ourselves, with their own Einsteins, and on the other hand there are people who dismiss any such intellectual achievement in the past, such as, for example, the knowledge of the length of a month or a year. It is therefore necessary to keep a fair balance between the two positions. To this end the participation of scholars from different research areas has to be achieved. Therefore any joint research between anthropologists, archaeologists, and astronomers should be welcomed.

Certainly the ancients did not use our modern tools and instruments. We may assume they had a lot of time and were innovative in their research procedures. We also assume that their astronomical knowledge cumulatively grew over hundreds of years, enabling them to perform particular observations with the use of rudimentary tools with great accuracy and precision. This fact is mentioned in order to point out one of the field's main limitations. Because of its materiality and the time depth an archaeological record can be transformed, preventing us from its exact reconstruction. It is thus impossible to establish the original precision and accuracy of the observations made by the ancients.

Brief history We think some brief introductory remarks to the field archaeoastronomy are useful in this place. As a result of Napoleon's expedition to Egypt many Europeans who worked there became struck by the precision with which the ancient monuments were made. Particularly remarkable were the pyramids with their four faces oriented to the cardinal points. Indeed the deviation of the pyramidal orientation from astronomical north is a very small one. The origins of archaeoastronomy may be placed in the latter half of the 19th century when the English astronomer Norman Lockyer and the German scholar Heinrich Nissen began their studies of the astronomical orientations of Egyptian temples and pyramids (Lockyer 1894) and Greek temples and Roman towns (Nissen 1906-1910).

Human beings and the stars: general considerations

Objects displaying regular and perfect behaviour, signs of a long-standing beauty, the stars seem to belong to transcendental beings that shine forever in the sky. Being placed in the sky by some mysterious and natural forces, they seem to move regularly in circular trajectories and have become symbols for perfection. Before being defined by spherical astronomy as ideal objects, the stars inspired the origins of astral mythology and eventually were transformed into complex symbols of universal sacredness, being considered as manifestations sometimes hostile, sometimes benevolent to men, divine powers. According to this view aspects of the life of individual persons depend on particularities of earthly environments and positions of the earth in a planetary system. All terrestrial processes are affected by the movements of the sun and the moon and other cosmic forces, which produce rhythms that affect geophysical, biological, and social phenomena. The universal rhythms originated in a physical world and produce biological and social rhythms on the earth, some of which are culturally encoded to form part of religious or cognitive models of the world. One of the basic manifestations of such rhythms is the annual seasonal round, which obviously affects human behaviour and patterns of activity. In sum there are obvious astronomical phenomena that have qualities of periodicity and duration. These annual rhythms are among the main forces that shaped human activities for a long time.

Elements of positional astronomy

A certain kind of proto-astronomy seems to be of concern to all peoples of all times. The regular sequence of the sunrise and the sunset, as well as the regular appearance of the full moon, could have become a counting device. The knowledge of the annual seasonal succession, or the appearance of the evening star, seems to be almost universal. We can assume, therefore, that this kind of knowledge is capable of becoming a sort of astronomy. Among the necessary conditions are: good conditions for observing the starry sky, the establishment of part-time specialists engaged in those observations, the development of recording devices enabling the transmission of knowledge acquired from the observations across generations, and last but not least, the justification of these activities as forming part of the official worship. Obviously such activities could have been fitted to agricultural or religious cycles, but on another level they could have been developed to foresee astronomical phenomena. In other words, in order to have an average level of astronomy, it seems necessary that there was knowledge of writing and some basic arithmetical system. The necessary conditions for the birth of astronomy appeared together in a limited number of places, Mesopotamia being chronologically first.

The studies of the orientations of ancient sanctuaries, or other structures related to worship, have shown that since the Neolithic onwards, on large areas of our continent and beyond the ocean, there existed a strict relationship between these structures and extreme positions of different astronomical bodies on the distant horizon. The name of the Dacians (or Getes) is mentioned by different Latin and Greek writers and refers to the peoples belonging to the northern branch of the Thracians. The present paper presents an attempt at reconstructing the ways in which the Dacians conceived the universe and, especially, how the astronomically determined great cycles formed the cognitive bases that determined much of their behaviour and organised their thought. As a consequence the study shows how the mental models representing those great cycles were systematically integrated in architectural structures and represented in the religious symbolism of the region in which the Dacians lived. Assumptions about the relationships between astronomical phenomena and material culture are to be reviewed from different disciplinary positions including anthropology, archaeology, astronomy, mathematics, archaeometry, archaeoastronomy, ethnoastronomy, and anthropoastronomy. As a result a kind of synthesis should address the major issues concerning the nature and role of the spiritual creations (science, myth, religion) of the Dacians. Practically, the present paper analyses alignments built upon orientations towards the significant positions of the sun, moon, and some stars on 'real' horizons. Some of the alignments encoded in architectural remains display a north-south orientation possibly associated with the idea of a local meridian; such seems be the case with several sanctuaries located at Sarmizegetusa-Regia, the ancient capital of Dacia, and other sites within the Orastie Mountains, e.g. Çostesti, Pustiosu, Brad, Fetele Albe, and Racos. The main sanctuaries of the sites mentioned usually consist of a central apse with an altar at one end and a series of columns supporting the whole construction. Our evidence indicates that the Dacians frequently encoded solstitial orientations in buildings. This may suggest that they were especially interested in keeping a record of those particular astronomical moments.

The Dacians: historical-social background

It can be shown that the Dacians were influenced throughout time by the cultures of the Scythes and Celts, but most of all by those of the Greeks and Romans. The old Dacian capital, Sarmizegetusa-Regia, was founded by Burebista (82-44 BC) who, with the assistance of the great priest Deceneu (Dicineus), unified all Dacian tribes under the same regime. As is known from the famous column of Trajan in Rome and from a monument at Dobrogea, the last Dacian king Decebalus (AD 87-106), who waged four wars against the Roman Empire, was finally defeated by the emperor Trajan in AD 106. After this victory Dacia became a Roman province and its former capital was destroyed and abandoned (Glodariu et al. 1988: 48-55).

Sarmizegetusa-Regia: orientations and archaeological data.

Direct access to Sarmizegetusa-Regia was protected by a series of defence fortresses, as at Costesti, Blidaru, Piatra Rosie, and Tilisca (Glodariu et al. 1988: 48-55). The remains of eleven rectangular and circular sanctuaries have been so far discovered in the area covered by the sacred precinct of the old Dacian capital. These were built at different times during the reign of Burebista (limestone constructions) and at the beginning of the second century AD, that is, during Decebalus' reign (andesite constructions). Although long known by local inhabitants, this impressive complex of ruins was not mentioned in the literature until the first years of the nineteenth century when the great hoard of Koson-Lisimachos golden coins was found.

We have looked at the structures erected on the tenth and eleventh terraces: the Great Limestone Sanctuary, consisting of four rows of fifteen plinths (Fig. 1); the Small Limestone Sanctuary, built with three rows of six plinths (Fig. 2); the Great Circular Sanctuary, forming a circle with a diameter of almost thirty meters and displaying a winter solstice sunrise orientation; the Small Circular Sanctuary of almost thirteen meters in diameter; two other rectangular sanctuaries made of andesite blocks aligned along the north-south axis; the Great Rectangular Andesite Sanctuary with six rows of ten elements each (Fig. 3), oriented towards the point where Sirius rose heliacally in the first century BC; and last but not least, an altar called 'Andesite Sun', a circular structure divided into ten sections and regarded as a type of sundial. Our measurements also indicate that the orientation of the axis of a sanctuary with horseshoe-shaped apse, located within a circles of andesite blocks and columns, marks the direction of sunrise at the winter solstice at that epoch. There are similar rectangular sanctuaries with apse at Çostesti (two large and three small limestone sanctuaries), Racos (Glodariu and Costea 1992: Fig. 5a, large circular sanctuary), Pustiosu, Meleia, Rudele, Fetele Albe, Dolinean (Moldova Republic), Brad, Barbosi-Galati, and Batca Doamnei, many of them displaying winter solstice sunrise orientations (Stanescu 1989; Schlosser and Cierny 1996: 101-102; Iwaniszewski 1998).

Astronomical knowledge of the Dacians.

The astronomical interest of the Dacians is mentioned by several ancient authors: Herodotus, Strabo, Jordanes, and others. The beginnings of their astronomical knowledge are attributed to Zamolxis (also Zalmoxe), considered as a historical person who was a religious reformer and legislator and a legendary king who was made divine. The first account is to be found in Herodotus (Godley 1982: 297 [Herodotus 4.95-96]): "For myself, I have been told by the Greeks who dwell beside the Hellespont and Pontus that this Salmoxis was a man who was once a slave in Samos, his master being Pythagoras, son of Mnesarkos ... but I think that he lived many years before Pythagoras". Strabo offers us more data. In his *Geography*, he writes, "it is said that a certain man of the Getae, Zamolxis by name, had been a slave to Pythagoras, and had learned some things about the heavenly bodies from him, as also certain other things from the Egyptians, for in his wanderings he had gone even as far as Egypt; and when he came on back to his home-land he was eagerly courted by the rulers and the people of the tribe, because he could make predictions from the celestial signs" (Jones 1924: 185 [Strabo 7.3.5]). We have similar information from Porphyry in his *Life of Pythagoras* (Nauck 1886: 19 [Porphyry 8-20]).

Fig. 1. The Great Limestone Sanctuary, called 'Burebista', is composed of four rows of fifteen plinths each. Another row of seven plinths has been added in the centre. The distance is 3.2 m between the rows and 2.0 m between the axes of the columns. Azimuth = -68° ± 0.5° along the short axis and +22° ± 0.5° along the long axis.

Fig. 2. The Small Limestone Rectangular Sanctuary at Sarmizegetusa-Regia. There are three rows of six plinths each. The distance is 4.30 m between the rows and 3.20 m between the axes of the columns. It displays the following alignments: azimuth = -68° ± 1° along the short axis and +22° ± 1° along the long axis. Calculations *in situ* suggest that the opposite orientation of the short axis marks the direction of the rising point of Sirius.

Fig. 3. The Great Rectangular Andesite Sanctuary. It consists of six rows of ten elements each and measures 35.0 m x 31.5 m. The plinths are 2 m in diameter; some of them are missing. Possibly the structure was left unfinished because of the war against the Romans. The azimuth of the short axis is -68° ± 1° and of the long axis +22° ± 1°. It therefore exhibits orientations identical with those of the two Limestone Sanctuaries. The opposite direction of the short axis is oriented towards the rising point of Sirius and its long axis to the rising of Capella.

Important for our knowledge concerning the figure of Zamolxis is Origen, according to whom it was not Pythagoras who introduced druidism to the Celts, but his disciple the Thracian Zamolxis, who also taught the Druids divination using coins and numbers (Miller 1851: 9). Among other ancient sources on the Dacians are the reports of Trajan or his doctors; unfortunately they are the last direct testimonials on the Dacians.

The most conclusive account of Dacian astronomy is found, however, in Jordanes, the Goth writer of the third century AD and author of the *Getica.* According to this work (Mierow 1908: 22 [Jordanes 11.69]), Deceneu, the above-mentioned great priest of the epoch of Burebista (1st century BC):

> ... taught them (the Dacians) almost the whole of philosophy, for he was a skilled master of this subject. ... By demonstrating theoretical knowledge he urged them to contemplate the twelve signs and the courses of the planets passing through them, and the whole of astronomy. He explained the names of the three hundred and forty-six stars and told through what signs in the arching vault of the heavens they glid swiftly from their rising to their setting.

Jordanes says furthermore (Mierow 1908: 23 [Jordanes 11.71]) that Deceneu

> ...chose from among them those that were at that time of noblest birth and superior wisdom and taught them theology, bidding them worship certain divinities and holy places. He gave the name of Pilleati to the priests he ordained.

This important passage tells us that the Dacians built sanctuaries for the worship of certain gods. Deceneu has been considered by generations of historians as a great reformer of Dacian religion. From this evidence it appears that from earliest times the Dacians paid much attention to astronomical events. The motions of the sun and the moon, the annual shift of the seasons, the rising and setting of stars, planets, and the sun were of considerable importance, affecting the course of human life. They were used to schedule agricultural activities as well as for navigational purposes. Whether in Europe or on other continents observations of the celestial bodies has proved to be a universal human phenomenon. Although these statements may be exaggerated, they cannot be entirely rejected because they seem to be supported by the archaeoastronomical investigations made by us.

Studies by Cantacuzino and Morinz (1963) at the Cernica Neolithic cemetery (4400-4200 BC) demonstrated that the great majority of graves (96%) were oriented within the two extremes of solar movement, suggesting thus that sunrise or sunset played a substantial role in establishing the orientations. The same rule for determining grave orientations survived until the Middle Ages. The observations made by Oproiu and Blajan (1990) imply that a similar proportion of the graves at the Alba Julia early Medieval cemetery, established around the 11th century, also follows this pattern (92%, or 167 out of 181). We can conclude that Dacian orientations were not exceptions in this regard.

Mircea Eliade, the well-known historian of religions, interprets the circular sanctuaries from Sarmizegetusa–Regia and Çostesti in terms of celestial symbolism, while the andesite altar called 'Andesite Sun' has been considered by him as other evidence for the so-called urano-solar character of the religious beliefs of the Dacians (Eliade 1972). We end with a quotation from Eliade (1972):

> There is another detail which also seems to be important to Strabo: Zalmoxis and later Deceneu were able to make such a prodigious career thanks to their knowledge of astronomy. The insistence on knowing the celestial movements may reflect the search for precise information. Therefore, with regard to current discussion concerning the existence or nonexistence of a common store of astronomical ideas of the Dacians, we maintain that they possessed a rather advanced astronomical knowledge.

The above-mentioned quotations are not the only ones that refer to the 'scientific preoccupations' of the Dacians. For example, according to Pârvan (1982: 14), a certain Dioscuride, a Greek doctor living during the 1st century AD, recorded a great number of Dacian names for medical plants. In this context it is worth mentioning that Romanian archaeologists discovered at Gradistea Munceluli the evidence for what is believed to be the remains of the house of a physician. The advanced level of Dacian technology is archaeologically confirmed by numerous finds of their weaponry, tools, and other objects used in everyday life; it is comparable to that achieved in the Graeco-Roman world (Glodariu and Iaroslavschi 1979).

Conclusions

1) The axes of the structures with an apse are oriented to, and mark precisely, the positions of the summer and winter solstices (i.e. $\pm \varepsilon$). The Great Rectangular Sanctuary and the Small Rectangular Sanctuary are aligned to the extremes of the moon (i.e. $\pm \varepsilon \pm i$). The existence of winter and summer solstice orientations occurs in different localities in the country (Sarmizegetusa-Regia, Racos, Pustiosu, Meleia, Fetele-Albe) and in different types of sanctuaries. In our opinion this fact clearly shows that the Dacian architects possessed advanced astronomical knowledge enabling them to orient their buildings and indicating that the buildings were oriented purposefully.

2) From the fact that some of these orientations cannot be established through a direct astronomical observation (because the views towards the North Star are blocked by the neighbouring skylines), it is deduced that the builders possessed either a sighting instrument or a developed theory that enabled them to determine those orientations. It is possible, therefore, that the structure called 'Andesite Sun', aligned upon the north-south axis, should be considered as a type of a sundial (Stanescu 1985).

3) The evidence also suggests that observations of solstices were of great interest. We infer, therefore, that solstices had a great significance for the Dacians.

These conclusions are drawn from an archaeoastronomical analysis of archaeological findings. Since we lack direct textual evidence of Dacian astronomy, archaeoastronomy appears to be a very useful approach.

*Lucian Blaga University, Faculty of Letters, History and Journalism, Victoriei Blvd 5-7, Sibiu 2400, Romania, smzfls@yahoo.com.

References

Aveni, A. F. 1985. *L'Archeoastronomia: scopi, ricerche, risultati*, In Primo seminario sulle ricerche archeoastronomiche in Italia. Brugine.

Cantacuzino, G. and Morinz, S. 1967. Die jungsteinzeitlichen Funde in Cernica. *Dacia* N.S. 7, 76-89.

Eliade, M. 1972. *Zamolxis the Vanishing God: Comparative studies in the religions and folklore of Dacia and Eastern Europe*. Chicago.

Glodariu, I. and Costea, F. 1992. Sanctuarul dela Racos, *Efemeride*, 34-39.

Glodariu, I., Iaroslavsche, E., and Rusu, A. 1988. *Cetati si asezari dacice in Muntii Orastiei*. Bucharest.

Glodariu, I. and Iaroslavsche, E. 1979. *Civilizația fierului la daci. (sec. II e.n.-I e.n.)*. Cluj-Napoca.

Godley, A. D. (tr.) 1982. *Herodotus*, vol. 2 (Loeb Classical Library 118). Cambridge MA and London.

Iwaniszewski, S. 1998. The development of a regional archaeoastronomy: The case of central-eastern Europe. In *Archeoastronomia, credenze e religioni nel mondo antico* (Atti Dei Convegni Lincei 141), 177-201.

Jones, H. L. (tr.) 1924. *The Geography of Strabo*, vol. 3 (Loeb Classical Library 182). Cambridge MA and London.

Lockyer, J. N. 1894. *The Dawn of astronomy*. London.

Mierow, C. C. (tr.) 1908. *Jordanes: The origin and deed of the Goths*. Princeton.

Miller, E. (ed.) 1851. *Origenis Philosophumena sive omnium haeresium refutatio*. Oxford.

Nauck, A. (ed.) 1886. *Porphyrii opuscula selecta*. Leipsig.

Nissen, H. 1906-1910. *Orientation. Studien zur Geshichte der Religion*, 3 vols. Berlin.

Opriou, T. and Blajan, M. 1990. Some results concerning the orientation of graves in Romanian prehistoric and ancient cemeteries. *Archaeometry in Romania* **2**, 35-40.

Pârvan, V. 1982. *Getica o protoistoirie a Daciei*. Bucharest.

Schlosser, W. and Cierny, J. 1996. *Sterne und Steine*. Dortmund.